The Alaska Gold Rush Letters and Photographs Of Leroy Stewart Townsend 1898 - 1899

Edited By

Peggy Jean Townsend

Patricia Roppel
(Alaska Historian)

Art Petersen
(University of Alaska Southeast)

CASTLE PEAK EDITIONS
Murphy, Oregon

by KLONDIKE RESEARCH
Auke Bay, Alaska

Thanks and Acknowledgments

To Blanche Moore Theiss for transcribing all of Leroy's letters.

To Arlene Treasure Townsend for working through many microfilmed newspapers to find articles about Leroy and the Keystone Company, for proofreading, and for encouragement.

To Charles W. Townsend III for encouragement, help in locating materials, and for laser printing a draft of the book.

To Robert Townsend Snowden (1929-1996) for lending nearly one hundred of the photographs taken by Leroy in Alaska.

To Edith Wallace Javens for lending pictures of Leroy.

To Vivian McLaughlin (1919-1997), director of the Resource and Research Center for Beaver County and Local History, Carnegie Free Library, Beaver Falls, Pennsylvania, for negotiating with the Pennsylvania State Library in obtaining local microfilmed newspapers of the 1897-1899 period.

To Larry Lashway for articles on Leroy Townsend from the *Seattle Times* and excerpts from Aberceombie's "1900" report to Washington, D.C.

Peggy Jean Townsend

FIRST EDITION (1000 copies)

Published by Klondike Research
PO Box 210174, Auke Bay, Alaska 99821
(907) 789-9830
In association with Castle Peak Editions
PO Box 277, Murphy, Oregon, 97533
(541) 846-6322 (74544.3010@Compuserve.com)

ISBN: 0-9657938-4-2 (pbk)   0-9657938-5-0 (cloth)

Library of Congress Catalog Number: 98-074544

Cover Photograph: "The Keystone Company," certificate of 2400 shares of capital stock, January 19, 1899, made out to Leroy S. Townsend. (Peggy Jean Townsend collection)
Rear Cover: Photograph of envelope from Orca, Alaska, dated APR 24 1898 (Peggy Jean Townsend collection), and envelope from Cordova, Alaska, dated DEC 03 1998.

Contents

Photographs and Maps

THE ALASKA GOLD RUSH LETTERS AND PHOTOGRAPHS

OF

LEROY S. TOWNSEND

1898-1899

PREFACE

by Art Petersen

LETTERS. When writing them, we usually do not intend to be historians. But we can be. Living in an historical context as we do, when we write about the physical and emotional events in our lives, yes, we address those who reside with us in that context, and usually for a practical purpose, not an historical one. But we can also address those who reside in the future. Such is the magical and wonderful power of words. They preserve our present, which inevitably must become the past, or *history*, if those words are preserved.

Photographs are as magical and wonderful, and Leroy Townsend in his letters often dwells on their power to preserve radiant moments in time, which he took great care to capture and to forward home as news of the immediate and retreating past. As the reader will see, Townsend was an historian at heart and in mind as he traveled rail, sea, land, and river roads to and through Alaska in 1898 and 1899.

Peggy Jean Townsend is also an historian and a very conscious one. She secured and now preserves the original letters and surviving photographs of Leroy S. Townsend, her father's first cousin. This substantive collection chronicles his participation in the Keystone Company in its fortune-seeking venture in Alaska. She came by the letters in 1969 and saw to their transcription in 1991. Subsequently, in 1993 she learned of the existence of Leroy Townsend's photographs, in the possession of Robert Townsend Snowden, and secured their use. In August of 1994 she made the collection available to Klondike Research for the purpose of publication. Her orientation to the project also bears the tenor of the historian: as she expressed it to me in a May 1993 letter, "My whole object is to make them [the letters and photographs] available for research and for perusal by those interested in the events."

In this same spirit, Patricia Roppel, an Alaska historian, was asked to study the collection, make observations about accuracy, write an historical context, and help gloss the letters. This work she accomplished by January of 1995, and in June of 1995, I visited with Peggy Jean at Lime Kiln Hollow Farm, a Townsend homestead (dating back to 1834), and together we studied a draft of the manuscript, checking questions against the original letters. In October of 1995 I visited Valdez to meet the landscape and confer with Joe Leahy, Director of the Valdez Museum and Historical Archive. We reviewed copies of the photographs, and I left him a copy of the manuscript for archive use. Joe later supplied interpretation of the manuscript, and so did Jim and Nancy Lethcoe of Prince William Sound Books and Jim Cook of North Pole, Alaska, a graduate student. Preserving the letters and photographs, verifying their content, giving them coherence, and providing historical interpretation have been a focused team effort contributed to by many.

With regard to historical interpretation, Peggy Jean Townsend has written an introduction and conclusion to the collection and provided many glosses throughout the work (signified by PJT). Patricia Roppel has provided an historical overview of the period, titled "Historical Perspective," and she also has glossed the letters extensively (signified by PR). Where it has seemed appropriate, I, as general editor, have sought appropriate placement of photographs and maps and their captions, including additional photographs and maps from three sources, the Patricia Roppel and Robert DeArmond collection, the archives of the University of Washington Libraries, and the archives of the Puget Sound Maritime Historical Society. I also added interpretive glosses where knowledge was in my possession (signified by AP).

The index is comprehensive. While working with Gold Rush literature, I have noticed how dense are the events of those times. With so many thousands of participants congregating in close proximities, many events are not only experienced by large numbers of persons; the persons themselves also interact with the lives of hundreds of others in the same locale. Their letters, diaries, photographs, legal documents, and books are evidence of how each person who recorded his or her experiences has touched the lives of so many others. Leroy Townsend's letters and photographs and the documents associated with his experiences (presented in four appendixes), list nearly two hundred persons who figure in that time, revealing their sicknesses, recoveries, and deaths; their tragic and pathetic circumstances; their brushes with disaster and their destruction by it; their noble and ignoble deeds. With so many names associated with Leroy Townsend's trek to the north, the only reasonable and considerate thing to do, it has seemed to me, is index these persons, in addition to other numerous references to such topics as wild game, locations, miners' meetings, prices, ships and shipwrecks, and storms.

The Townsend collection is a valuable one. It does not provide special new historical information about the period. It does not even reveal anything new about the scurvy epidemic that became so much a part of the Valdez and Copper River Valley experience of 1898 and 1899 (between November 9 and April 5, no Townsend letters are known to survive). The collection is valuable, however, for its portrayal of the hopeful and youthful spirit of a thoughtful and generous man of character. Leroy Townsend is an observant and often thoughtful and graceful recorder of events both around and within him. Thus does he add in a substantive, interesting, and generally engrossing way to the human coloring of the time. This collection is also valuable for the photographs he took during his sojourn. But the collection is also valuable in another way.

The absence of letters during the scurvy-ridden winter of 1899 is partially due, no doubt, to their loss in transit and to the isolation of the Copper River region in that pioneer year, but it also seems likely that letters or a journal was not penned because Townsend was consumed by duty to dozens of ailing men. That he saved the lives of many is attested to by published documents (journals, letters, military reports, and newspapers). In following his Hippocratic Oath as a medical doctor, without deviation or stint, Leroy Townsend distinguished himself above the some twenty-five doctors in the region that winter. He deserves the tribute this book brings to his diligence as a person among so many others in difficult circumstances, as an historian of his time, and as one who is distinguished in his dedication to life-saving efforts.

History has a way of rewarding people for this kind of service. Beyond this book is another reward, "Townsend Peak" (17 miles north of Valdez). It was named in 1960 by L. E. Nielsen for service rendered desperately sick men during the winter of 1899 (Orth 1967). Townsend Peak overlooks the Valdez and Klutina Glaciers, which Townsend traversed many times, the last of which was with 5 scurvy patients. They had sledded 82 miles from Copper Center when Townsend called a halt at the foot of Klutina Glacier. Quartering the sick in a cabin at Barrett's Camp, Townsend and 3 others set out for help 17 miles away over the stormbound glaciers.

Scurvy patients being sledded from Copper Center toward Valdez, April 1899: "We left Copper Center, Monday, April 17, at 5 a.m. We succeeded in getting the assistance of three Indians and had the help of three men from the Teikel. This, with the hospital corps of four, made the relief party number ten. We were sledding five patients [Schloegel, Hunt, Launert, Liljegren, McCracken] and duffel, with provision, camp equipage, etc.; so you will readily appreciate that our task was no light one" (Ltr 29). (Snowden Col., Townsend, April 1899)

They pressed on past the point where three of the Logan Party of 6 had frozen to death, and Townsend and the others almost stumbled into the same area of crevasses into which 2 others of the Logan Party had fallen and died. After a night on the Glacier under mounting snow, the party made the Dempsey Relief station the next day, with Townsend too tired and sore to continue. For one who, as testified to in his letters, prided himself on his packing ability and prowess under difficult conditions, this decision reveals the arduousness of the trek. "I thus became a case of a relief party relieved," Townsend writes (Ltr 29), when a party of two arrived with dogs and a sled, sent by Quartermaster Brown. Townsend allowed the men to sled him to the third bench but insisted on walking the rest of the way to Valdez. Then a party of 9 men sent by Quartermaster Brown successfully extracted the 5 sick men from the interior (Brown, Appendix A).

Townsend Peak oversaw this close of Townsend's winter of ministrations in the Copper River Valley, an enduring 5700-foot monument to a man who prevailed over many trials and to the travails of thousands of men and some women during the years 1898 and 1899.

As general editor of this collection, I have followed the precept that always seems best: to pursue and preserve the intention of the author in an effort to gain "achieved content" while also seeking a readable, clear text and format. In this regard, some precepts of standard punctuation have been applied in the pursuit of clarity. Some standard grammar and spelling have been supplied, but very little. Some spellings (e.g., *Valdes* and *Pittsburg*) have been preserved as appropriate at the time and to its atmosphere.

TRUTH, FALSEHOOD, AND THE MIDDLE GROUND. Recently I came across an interesting perspective about photographs, expressed by Michael Lacy. In 1973 he published a controversial collection titled *Wisconsin Death Trip*. Of this collection and others of his, he is recently quoted as saying,

> Academics hate photos because they hate paradoxes. They like linear arguments. What photos can do—if you let them—is to move toward truth and falsehood at the same time. (*Lingua Franca*, March 1998, 29)

This perspective seems not only applicable to photography but also to other mediums, including the arts at large as well as letters and documents. What Lacy's statement illuminates, I think, is part of the addictive appeal of the humanities, *history* and all that makes it. In history, including its enticing ambiguities, resides a complex, rich, and illuminating perspective on the nature of our being. Each of us is enriched by our association with the subtle nuances of the history of our kind. And here for your enrichment, as well as pleasure, is another such association, the middle ground of which you are invited to find for yourself.

Art Petersen
Bozeman, Montana
July 1998

INTRODUCTION

by Peggy Jean Townsend

THE THIRTY LETTERS that form this collection were written near the end of the nineteenth century by Leroy Stewart Townsend, a medical doctor from Beaver Falls, Pennsylvania. In 1898, he joined others under the direction of Colonel H. R. Creighton in the Keystone Company, which was formed for the purpose of seeking gold in what has come to be known as the Copper River Valley, Alaska. According to an agreement (Appendix E), Dr. Townsend was grubstaked by his father, L. G. Townsend, and his uncle, C. W. Townsend, in return for 5/12s of a one-half interest in the proceeds from any claims.

During his eighteen-month adventure, Dr. Townsend faithfully wrote letters to his family and took photographs along the way. The letters were found among the papers of his sister, Elizabeth Adams Townsend (1879-1969). All were addressed to his father "L. G. Townsend, Agent Penn'a Co., Beaver Falls, Beaver Co, Penn'a."

Dr. Townsend was twenty-seven when he embarked for the north. He was born April 8, 1871, and died August 24, 1936. He was educated in Beaver Falls schools and for a year attended classes at Geneva College. After teaching briefly at McMahon Business College in Beaver Falls, he enrolled in medical studies at Western University, now the University of Pittsburgh, and finished training with post-graduate work at the Polytechnic University of Philadelphia.

In 1896, Dr. Townsend returned to Beaver Falls to start his medical practice, but the glamour of the Alaska gold fields called him, and he became a member of the Keystone Company, one of a multitude of companies incorporated to prospect in the north country. With the discovery of gold in the Klondike and its first report in July of 1897, newspapers hastened to print stories about the gold seekers. The Keystone Company's plans and progress were duly reported:

[Beaver Falls *Weekly Tribune,* January 26, 1898]

PREPARING TO LEAVE

Members of Col. H. R. Creighton's Company to Leave for Alaska on Thursday. Organization Formed Yesterday.

The parties who are to accompany Col. H. R. Creighton, of California, to the Copper River in Alaska for the purpose of prospecting in the gold fields at that place met at the office of attorneys McClung & Evans in Pittsburg yesterday for the purpose of signing articles of agreement and forming an organization.

The persons who signed the articles were Col. H. R. Creighton of California; Rev. O. A. Richards[1] and son H. W. Richards; and Dr. Roy Townsend of this place; Messrs. T.

[1] *O. A. Richards*: Rev. Orphanus A. Richards, minister of the Church of Christ in Beaver Falls; he is given a nine-month leave of absence to seek "health [indigestion] and gold" (*Christian Standard*, 2/19/1898, 236).

Sharp and son and Wilber Henning of Braddock; a Mr. Scott of Chicago; and Mr. Bruce of Pittsburg. W. P. Barnum and B. W. Rayl of this place expected to complete arrangements so that they could sign the agreement today and accompany the excursion which will be joined at Seattle by several persons from the Pacific coast. The officers elected were: President and general manager, H. R. Creighton; vice president, O. A. Richards; treasurer, Wilbert R. Henning; secretary, Mr. Bruce.

The gentlemen expect to complete arrangements for transportation to Seattle so that they will be able to leave by eight o'clock on Thursday morning.

Opposite page: The 12 members of the Keystone Company in Seattle prior to sailing for Valdez. From R to L, Leroy Townsend stands 6th, holding camera. This photograph is by Ashahel Curtis, a commercial photographer who captured many striking pictures of the Klondike Gold Rush. Rev. Orphanus Richards is believed to stand at the L, with Col. Creighton, president, to his immediate right. All appear to wear new boots, many with bright leather laces. (Snowden Col., Curtis, #394, Feb 1898)

[Beaver Falls *Weekly Tribune,*
February 2, 1898]

LEFT FOR ALASKA

Col. H. R. Creighton's party of nine persons left Saturday for the gold fields on the Copper River in Alaska. Dr. LeRoy Townsend left Friday night for Chicago and will join the party at that place. Rev. O. A. Richards and son Hiram and B. W. Rayl boarded the train and joined the party at this place. W. P. Barnum, who had made preparations to become a member of the party, was unable to complete arrangements for leaving his business here in time to leave to-day and was compelled to change his plans. Mr. Barnum with a number of relatives and friends of the members of the party was at the depot to wish the gold seekers success.

In March of 1898 the *Weekly Tribune* published a letter from O. A Richards, who gave an account of the party's progress.

[Beaver Falls *Weekly Tribune,*
March 30, 1898]

LOCATED IN ALASKA

Colonel Creighton's Party Crossing the Glacier to Copper River. A Letter from Rev. O. A. Richards States that the Company Has Located a Trading Station at Port Valdes and Were Ready to Begin the Work of Crossing the Mountain of Ice to Their Destination – The Members of the Party All in Good Health and Excellent Spirits.

Rev. O. A. Richards, who with several other residents of this place, left several weeks ago with Colonel Creighton's party to locate in the Copper River district of Alaska to prospect for gold, writes from Fort Valdes, Alaska, under date of March 3.

He gives an interesting description of the trip from Seattle, Washington, on board a steamer [the *Alliance*] upon which there were two hundred and fifty passengers and about fifty dogs bound for the Copper River district. During the voyage the ship encountered considerable rough weather, and all but two members of the company were very ill from seasickness, the decks of the ship being drenched repeatedly with spray, which the chilly wind froze into ice.

He described the beautiful sights obtained of glaciers extending from two hundred to three hundred feet above the surface of the water and to the depth of seven hundred feet beneath the surface.

The town of Juneau he describes as a little nook in the mountains, which tower for hundreds of feet almost over the town, looking as though a snow-slide might come down at any time and cover the entire place. He also described the grandeur of Mount St. Elias and Mount Logan, which were visible 40 miles from shore and were passed on February 24.

Juneau, circa 1897-98. (Special Collections Division, Taber, UW Libraries, #: UW 18283)

The country about Orca [present-day Cordova], which was reached on February 25, is described as far from being the barren, forsaken country it had been represented. On arriving at Port Valdes on February 30th, the steamer was compelled to anchor to the ice three miles from shore, where their goods were unloaded onto the ice. After being sorted the goods were placed on sleds, and the work of hauling the effects to land was commenced and continued until completed, as there was some danger of the ice breaking loose from the shore and floating out to sea with the men and goods. Rev. Richards and the other members of the party each made five trips to the shore, covering altogether a distance of about thirty-five miles with their sleds before the work was completed. Tents were erected and the members of the party spent the first night in Alaska rolled up in their sleeping bags in the tents.

The next work was the cutting through sixteen feet of snow and ice to secure a ground foundation for a frame building erected for a store-house and trading post. The members of the company were favored with good weather and made good progress. Richards describes the halibut and cod fish taken from the bay as being the finest fish he ever ate, and as they were easily caught, the company was well provided for in that line.

On Sunday the men took a well earned rest, as all were sore and tired from the unusually hard work they had performed. On that day the steamer *Rival* arrived and discharged one hundred and seventy-five passengers and one mule at that place, which made a colony of four hundred men and seven women at the end of the bay where the company had located.

At some places the slope is so steep that the sleds have to be taken up with blocks and tackle, which makes the work slow and tedious. The members of the company expect to commence on Tuesday, March 8, to move their goods to the summit of the glacier, after which they expect to reach the Copper River valley with very little difficulty providing the melting ice does not cut furrows so deeply in the ice that they cannot be crossed. The furrows are said to be at times fifty feet in width and one hundred feet in depth. The company has located one hundred and sixty acres of land on the shore of the bay at Valdes where they have established a trading station and anticipate doing considerable business this season as a number of improvements are anticipated in that vicinity.

In writing regarding persons going to that country, Rev. Richards advised the organization of strong companies who should have plenty of supplies and men enough to be independent. He thinks the many persons going over the route to the Copper river will have an open trail constructed by next winter; then the greatest danger will be from the severe cold weather, which is often accompanied by a wind so strong that men are often blown from their feet and carried along the ice. He also advises keeping a sharp watch on the merchants at Seattle, who will fill your order with second rate goods unless they are looked after while the goods are being packed.

The Pacific Steamer Whaling Company have a tract of land adjoining Creighton's upon which they are preparing to erect a large hotel, they having the lumber for that purpose already in the bay.

Honesty is proverbial in that country, as a man who steals less than $100 is treated to the whipping post, and one who steals over that amount is punished by death. The writer speaks of seeing a man put up a notice that he had found a jack knife along the trail. The posting of a notice announcing the loss of an article will usually cause the missing article to be returned.

Many men who have attempted to cross the glacier have become discouraged, sold out and are returning home; but the members of the Creighton party are all enjoying good health and determined to remain and succeed.

As Leroy Townsend's letters testify, the optimism of most members of the Creighton party held up during the months that followed despite the extreme difficulties encountered both in crossing the Valdez Glacier and in their search for gold in paying amounts after the crossing. As prospects slimmed in one prospecting expedition after another, however, members of the party began resigning from the company or looking beyond the Copper River Valley for better prospects.

The thirty letters in the collection may be divided into eight locales, and to present them as a chart provides a clarifying overview to the sequence of events they cover:

General Locale	Ltr.	Date	Days Since Last Ltr.	General Locale	Ltr.	Date	Days Since Last Ltr.	General Locale	Ltr.	Date	Days Since Last Ltr.
1. Seattle	1	2/3/98		**Area**	11	3/15/98	9	**Klutina L.**	21	6/13/98	6
"	2	2/10/98	7	**Glacier**	12	3/20/98	5	**6. Amy's**	22	6/28/98	15
"	3	2/15/98	5	"	13	3/30/98	10	**Landing**	23	7/29/98	31
2. S.S.	4	2/19/98	4	"	14	4/14/98	15	"	24	10/10/98	73
Alliance	5	2/21/98	2	**4. Timber**	15	4/22/98	8	"	25	10/10/98	0
"	6	2/23/98	2	**Beyond**	16	4/29/98	7	"	26	11/9/98	30
"	7	2/25/98	2	**Glacier**	17	5/8/98	9	**7. Copper**	27	4/3/99	145
"	8	2/25/98	0	**5. Trail**	18	5/15/98	7	**Center**	28	4/13/99	10
"	9	3/4/98	7	**Valdez**	19	5/26/98	11	**8. Valdez**	29	5/2/99	19
3. Valdez	10	3/6/98	2	"	20	6/7/98	12	"	30	5/7/99	5

June and then September and October of 1898 were busy months, full of the activities of transporting goods and prospecting unknown, difficult terrain, and then later, hunting in preparation for winter. Leroy Townsend had hunting experience and a facility with firearms, and the tenor of his letters during this period is one of relief as he threw himself into a familiar activity. The letters convey a feeling of comfort in the Alaska environment, of wonder over its expanse and bounty, and of pleasure in the company of men who had prevailed in reaching the interior with their outfits relatively intact and who providentially prepared to spend the winter in the Copper River Valley.

The only known letter written in November of 1898, and the last for 145 days, was begun November 9, but it was not posted until January 24, 1899. Probably mail service had become sporadic, for it was in this November letter that Leroy reports having received a letter from home, dated August 26. It seems likely the November letter was posted in December as we have a photograph documenting a later event, a "Thanksgiving Dinner."

So opportunity to forward mail could have been sparse, and we know that Leroy was undertaking treks with others for fifteen days at a time, prospecting for sites to be followed up on in the spring and hunting. Not known during this period is the degree to which men were beginning to show signs of scurvy nor the degree to which Leroy was called upon to render medical attention. Already a member of the Keystone Company, Hiram Richards, had been convinced by Leroy to return home as Richards was showing signs of scurvy. Then, as Leroy puts it in a July 1899 interview with the Beaver Falls *Weekly Tribune*,

> We ... were ready to go into winter quarters when scurvy broke out at Copper Center, and they sent for me.... There were 125 men sick at one time, and others coming in daily. I had thirty helpless men on my hands at one time.

For whatever reason, we have large gaps in the correspondence, but they are filled in somewhat by the research of Patricia Roppel and by Appendixes A, B, and C: the military Quartermaster's Report, the report on scurvy by Leroy Townsend, and his request for remuneration.

Following Patricia Roppel's interesting Historical Perspective and then the letters of Leroy Townsend, I will share with you in a Conclusion the end of his northern adventure and something of what occurred in his life thereafter.

✸

HISTORICAL PERSPECTIVE

The Copper River/Valdez Glacier Route To the Klondike

by Patricia Roppel

ON AUGUST 16, 1896, George Washington Carmack discovered gold on Bonanza Creek in Canada's Klondike. Thus began the greatest gold rush of northern history. Almost as fast as news of the gold strikes spread across the world, advice to would-be prospectors began to appear in print. Although the Copper River, which empties into Prince William Sound, Alaska, had seldom been successfully ascended, and then only with great difficulty in small boats, some "experts" touted it as a route to the gold fields.

The Copper River, known to the aboriginal inhabitants as the Atna, begins in the Wrangell Mountains, which rise on the Alaska-Canada border in southcentral Alaska. Starting as the melt waters of the Copper Glacier, which hangs on the north face of 14,000-foot Mt. Wrangell, the Copper River descends 287 miles along the foothills of these mountains, then enters the Chugach Mountains where it becomes very narrow before merging with the flats of the delta as it flows into the Gulf of Alaska about 27 miles southwest of present-day Cordova.

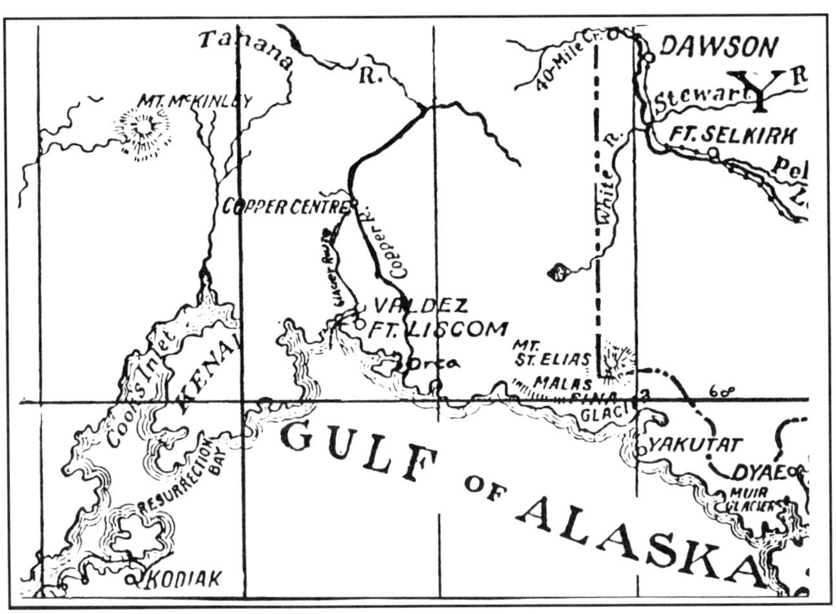

Map showing location of Orca, now Cordova, at the mouth of the Copper River, and Valdez to the NW.

While the Copper River appears an inviting and broad corridor to interior Alaska, it proves formidable. Its gradient produces a swift current, averaging seven miles per hour. The total fall of the river between its head and the sea is approximately 3,600 feet, for an average drop of 12 feet per mile. On the lower half of the river, from Copper Center to the mouth, it falls about 7 feet per mile. This is a very steep gradient compared to that of the Yukon River, which falls 1.2 feet per mile from Whitehorse to its mouth on the Bering Sea (Mendenhall 1905, 20).

Prior to the appearance of gold seekers, a few persistent traders, explorers, and prospectors ascended the Copper River. Before their coming, it was the land of an indigenous people. Here in the Copper River Valley from the river's headwaters to Wood Canyon was the land of the Ahtna, interior Athapaskans. A no-man's land stretched from the canyon to Childs Glacier, the northward limit of the Eyak Indians who lived on the Copper River delta in Prince William Sound.

A Russian trapper in 1796, scouting for fur-bearing animals, was reportedly the first European to see the Copper River and ascend to the valley above (Hanable 1982, 19). Over the following years, a number of parties explored upriver, and their first-hand recorded knowledge of the river included many significant facts such as the river's location, general course and characteristics, and resources. However, little of this information was readily available at the end of the nineteenth century to gold seekers.

Captain (here Lieutenant) Abercrombie (U.S. Army, Fort Richardson, Alaska. Roppel-DeArmond Col., circa 1884)

More accessible, but generally ignored by Klondikers, was the account of Army Lieutenant William R. Abercrombie's failed 1884 attempt to explore the Copper River. After Abercrombie and his party turned around near Miles Glacier, Natives informed him that an abandoned Indian trail started at the head of Port Valdez and went over the mountains to a lake, the outlet of which supposedly ran into the Copper River above the canyons and rapids. Abercrombie set out to cross the Valdez Glacier in mid-September, but he found the snow too soft to proceed.

The next year, 1885, another Army officer, Henry Tureman Allen, successfully explored the Copper and the Chitina rivers. After negotiating difficult portages, sledding in soft snow, and boating in treacherous water, he reached the main valley of the Copper River and carried his explorations across the divide to the Tanana River by way of Mentasta Pass, then to the Yukon River and down it to St. Michael on the Bering Sea coast. His travels set the stage for the future of the Copper River corridor to be seen as an all-American trail to the Klondike gold fields.

The gold seekers of 1897-1898 did not find accurate information about routes to the Klondike because the government was unprepared for the flood of inquiries about Alaska. The supply of U. S. Coast Survey maps was soon exhausted, and the demand for information became so great that newspapers began printing the advice of returning miners and others who claimed first-hand knowledge of the country.

In the frenzy of gold fever, armed with optimistic advice of dubious reliability and after disregarding somber counsel, men and some women hurried north. Indeed, few questioned the report of a "new route" to the Klondike by way of the Copper River. The Valdez Glacier route, heavily advertised in the United States as the "All-American" route, avoided the galling and expensive Canadian import duties[1] on outfits as the gold seekers entered the Klondike by the Skagway/Dyea access at the head of Lynn Canal.

In addition, financially-interested parties often misrepresented conditions. One such entity was the Pacific Steam Whaling Company as represented by Captain O. J. Humphrey. An employee at the company's salmon cannery at Orca near the mouth of the Copper River since 1889, Humphrey was readily quoted in San Francisco and Seattle newspapers especially after he arrived in San Francisco with specimens of quartz reportedly containing $20 in gold and $10 in copper per ton.

Humphrey correctly emphasized that attempts to ascend the Copper River would be futile, but he pointed out that travel over the Valdez Glacier would take prospectors to the Copper River and Klondike regions (Hanable 1982, 50). Within two weeks of Humphrey's statements, the Pacific Steam Whaling Company inaugurated passenger transport to Alaska. On November 10, 1897, the company's small steamer *Wolcott* sailed from San Francisco for Prince William Sound. Soon after, the company added three more steamers to the run. It did not take long for other steamship companies from Portland, Seattle, and San Francisco to enter the field.

In 1897, a few parties prepared for winter travel into the Copper River. Even after many of these disappointed stampeders returned to Pacific Coast ports with stories discrediting the

---

[1]*Canadian import duties*: In early February of 1898, the Northwest Mounted Police required each Klondiker to possess a minimum of 1,150 pounds of food as well as other necessities amounting to 2,000 pounds or be denied entrance to the Territory (Burton 1958, 165, *Dawson City* 1988, 27). The March 1898 issues of *Klondike Grubstakes: Practical Information for Yukoners and Others* contained the "Complete Canadian Customs Regulations" for import paper work, processing, and duties. The following is a sample:

> The person entering any goods inward shall deliver to the collector, or other proper officer, an invoice of such goods, showing the place and date of purchase and the name and style of firm or person from whom the goods were purchased, and a full description thereof in detail, giving the quantity and value of each kind of goods so imported, and a bill of entry thereof in such form as is appointed by competent authority, fairly written or printed, or partly written and partly printed, and in duplicate, containing the name of the importer, and if imported by water, the name of the vessel and of the master, and of the placer to which bound, and of the place within the port, where the goods are to be unladed, and a description of the goods and contents of the packages and the place from which the goods were imported, and the country of which the goods are the growth, produce or manufacture. (Cohen 1996, 9)

As for the duties, they ran from 10% to 35%. Then there were the royalties on minerals found, established in 1898 at 10% (Haskell 1998, 441).

boomers, the rush of would-be gold miners up the Copper River Valley continued during the winter of 1897-1898. About 300 tried to go directly up the Copper River from tide water. Most did not get past the Bremner River. Nearly 4,000 gold seekers, including Dr. Leroy Townsend, chose the route abandoned by Abercrombie over the Valdez Glacier. The route was not limited to men: the tide of gold seekers included an unmarried woman as well as many women who accompanied their husbands.

Few who arrived at the head of Port Valdez had any idea of what lay ahead. The route, they soon found, involved a difficult and often treacherous journey over the Valdez Glacier, which fills a pass through the rugged Chugach Mountains. Once safely past the crevasses and endless ridges on the seaward side, the gold seekers planned to descend the Klutina Glacier, build a raft or boat to negotiate the braided channels of the upper Klutina River, sail Klutina Lake, and descend the rapid-filled lower Klutina River to the Copper River.

The first prospectors to negotiate this route established the camp of Copper Center on a low terrace at the junction of the Klutina and Copper rivers, and it became the headquarters for travel up and down the Copper River. From here some gold seekers planned to cross the river at Copper Crossing and head for the Millard Trail, which had been laid out by members of the Wisconsin Company and Great Northern Party. That trail ran north past Mt. Drum and Mt. Sanford to the mouth of the Slana River. Here the men followed the eastern bank of the stream, which led to Mentasta Pass through the Tanana Mountains, and on the north side they expected to strike Fortymile River, which they could follow down the Yukon River to Circle City.

Some headed for the Klondike via this route, but the majority, such as Creighton and members of the Keystone Company, intended to prospect the little-known Copper River Valley. They were lured by vague rumors of mineral wealth in the interior of Alaska and hoped to find placer gold in the Copper River tributaries where there would be less competition.

In many ways the "All American," Valdez Glacier route was the most difficult and hazardous route to Alaska's interior and the Klondike gold fields. Unlike the White Pass and Chilkoot routes at the head of Lynn Canal in Southeast Alaska, where most of the 600-mile journey to the Klondike was by river, the stampeders who entered Alaska by way of Valdez spent most of their time handling their goods and traveling by one means and then another (sled, pulley, boat, and foot). As for hazards, Horace Conger, who made the trip, mentions chest pain, stroke, smashed fingers, an accidental shooting, a broken leg, sore throats, and a hanging that resulted from a miners' meeting—all before he started over the glacier (Holeski 1983). Once on the glacier, he and others suffered from frostbite, immersion foot, snow blindness, scurvy, and neuralgia of the jaw, not to mention the injuries resulting from snowslides and falls into deep crevasses, which were only loosely concealed by snow in early winter before heavy snows and freezing winds formed solid bridges. However, there was never a shortage of doctors; at one time as many as twenty-five physicians, including Townsend, were on the Valdez trail. Because of the hardships, men banded into parties such as Townsend's Keystone Company. Many men (including Townsend) and groups were "grubstaked" by men who could not go to the Klondike but wanted a piece of the action if gold were discovered in the north country.

From Valdez to the summit of the glacier, the miners believed, was 23 miles, just a little farther than the distance from Dyea to the summit of the Chilkoot Pass, and for many, it took the same time to negotiate the route, about two months of slow, back-breaking toil. Later a geologist determined the distance actually to be 19 miles. The gold seekers made the trip to the summit in stages described by the shape of the glacier. A trail to the foot of the glacier led in nearly a straight line from the beach at Valdez across the large flat. This was a gentle rise of only 150 feet for 4 1/2 miles. Here the men encountered the first bench. Benches are changes in elevation on the glacier, and most were steep inclines. Five benches marked the slope to the 4,800-foot summit of the Valdez Glacier. The tops of the benches were terraces of varying sizes. The first three were located near the foot of the glacier. From the first bench to the second was about a mile long with a gradual grade of 55 feet to the foot. The top of the second bench was only 200 yards to the foot of the third bench, which measured 60 feet; it started with three sharp pitches, each with a grade of 15 degrees.

Looking toward Valdez, likely above the first bench (Roppel-DeArmond Col., # 136, winter 1898)

Then the really hard work began. The distance from the top of the third bench to the foot of the fourth was about 8 miles with a rise in elevation of 1,920 feet. The fourth bench was 6 miles in length and gradually climbed 1,050 feet. The fifth bench was also known as the "foot of the summit." The slope to the summit was about a mile, and over this distance the elevation was about 1,000 feet.

Prospectors as a rule carried an outfit averaging 1,500 pounds per man. The main means of transport were sleds measuring about 18 inches wide and 6 feet long. Few prospectors had pack animals, although horses, mules, burros (which proved unsatisfactory as their hooves were so small they broke through the snow's crust), dogs (often Newfoundlands or St. Bernards), and even goats were seen on the trail. A good load for a man, even on the gradual slope of the glacier, was only 150 pounds, according to one army officer (Glenn 1899, 398). Therefore, an outfit of 1,500 pounds necessitated at least 10 trips over the same ground. A good, strong worker could make the summit with 1,500 pounds in 18 days of good weather, of which there was very little.

At certain places, such as the faces of benches, block and tackle were required to hoist supplies up the steep inclines. One gold seeker, C. H. Remington, describes this process: first a "dead man," or stake, was anchored in the ice a distance up the slope equal to the length of the rope upon which a single block pulley was fastened (Remington 1939, 10-11). About 300 feet of three-quarter-inch rope with one four- or six-inch single block fastened to the "dead man" was best, making for a 150-foot haul; a much longer rope proved harder to drag. One end of the rope was fastened to a loaded sled at the bottom of the incline, and with one group of men pushing up on the loaded sled, the other group, holding the other end of the rope, ran down the incline, forcing the loaded sled up. Then the latter gang, backpacking additional supplies, reclimbed the steep incline. And the process began all over again. About 6 men were needed to drag up a load of 300 pounds.

When not using block and tackle, men loaded their sleds with as much as they could pull, and with men hauling on lines ahead and others pushing from behind, they worked their way upward for one-to-two miles before making a new camp. Then back they went to the old camp for another load, relaying four or five loads a day. The process would be repeated day after day until the entire camp with all the provisions and supplies was moved to a new site. In addition, some men carried oil and a stove, or as most men did, sufficient wood to last the trip on the glacier. With dehydration a major problem, fuel to melt ice and snow for coffee or tea was necessary for survival. Food could be eaten cold. Men could huddle in their tents, wrapped in sleeping bags for warmth, but at the end of a sweat-slogging day, bodily fluids had to be replenished.

The elements did everything they could to make the trip unpleasant. There was usually a consistent wind, and it was sometimes accompanied by gusts of sleet and snow that coated men and beasts with an armor of ice. More than one man was lifted clear of the ice and carried several yards by the force of the wind. Temperatures were reported to reach 48 degrees below zero. The trail disappeared under fresh snow fall, and sometimes during a storm, it became the custom to keep men traveling over the trail with empty sleds or with snowshoes to keep it open. However, when visibility was reduced to only a few feet during a blizzard, the gold seekers were immobilized in camp where they remained in their sleeping bags, emerging only long enough to cook when they were hungry. These delays were frustrating as the men consumed rations that cost so much labor to transport. More days would be lost as men dug the equipment out of the fresh snow to prepare for another move. No wonder Townsend speaks with great satisfaction when his party purchased supplies already packed up the trail!

Camp below the summit, the fifth bench, and the summit beyond. A testament to strenuous working conditions lies on the packer's sled, his coat, to the right of which is a large flask—of water, presumably. To the packer's left is a dog residing on supplies, and to its left another dog relieves itself. A man looks in the direction of the photographer from behind a large pile of supplies, and to his left is another dog atop supplies. The camp is seen vaguely further on, with the trace of the trail to the summit mounting steeply back to the right. (No. 25, Roppel-DeArmond Col., winter 1898)

Leroy Townsend on snowshoes, probably on the Valdez Glacier, February 1898, in nearly "white-out" conditions. Note the heavily loaded sled right. Resting on Townsend's snowshoe is the case for his camera, probably handed to a Keystone Company companion for this photograph. (Snowden Col., winter 1898)

When Townsend crossed the glacier in April 1898, it was estimated that at least 700 people were on the trail (Glenn 1899, 300). Many of the gold seekers, including his party, worked at night when the cold froze the upper layers of snow into a solid crust, which made travel easier. The constant travel of one thousand people going back and forth in March and April packed the trail, which was marked by stone monuments that tended to slide out of position as the ice melted on warm days. As the warmer days of April came, the trail began to disintegrate on the lower benches, becoming slushy after 9 a.m. but stiffening up enough to bear a man by about 9 p.m. and enough to bear horses by 1:00 in the morning.

Avalanches and small snowslides upon the mountainsides of the glacial gorge were continual, but people were impatient to move ahead for fear the glacier trail would become impassable. Townsend's party crossed the glacier before the disastrous blizzard and avalanches that took place at the end of April, but he mentions the event in two of his letters. Although few people were killed, the *Seattle Post Intelligencer*, May 20, 1898, likened the snowslide to the terrible Chilkoot disaster.[2] The snow began to fall on April 28, and after four days, 5-to-8 feet of heavy, wet snow had fallen before a warm rain began. The blizzard immobilized everyone. The first snowslide occurred the evening of April 29, bringing with it a large quantity of rock and causing a roar that could be heard for five or six miles. This slide took place on the third bench and extended over the trail, burying six men who had camped there at the beginning of the storm, along with some of their 25 burros. The men were all taken out alive, suffering from severe bruises, but ten burros were killed and the remainder badly battered (Glenn 1899, 397).

On the night of April 30, tons and tons of fresh snow, the first of a series of slides further up the glacier, tumbled from the summit and spread a smothering blanket over the middle portion of a long line of tents that 400 men and women had pitched at the foot of the glacier. Somewhere between 25 and 60 people (depending upon whose version of the event) were dug out of 10 feet of snow within an hour of the slide. They were located by their cries, plainly heard through the snow, and all but two were taken out alive. These men were under a heavy tent that, upon being covered by the slide, had smothered them. The dead, according to a Seattle paper, were Joseph Furner (or Fournier) of Chicago and B. Van Antwerp of Minneapolis. A. Johnson of Eureka, California, was caught on the edges of the slide and received injuries from which he died.[3]

The Seattle newspaper account lists another slide on the following night, which swept down from the summit and struck the north end of the line of tents. Again the work of rescue commenced, and in a few hours forty half-smothered, half-frozen men and women were taken out. Part of the slide was not explored, although it was believed that from 15 to 20 occupants were still covered. It was impossible to reach them that night, and a smaller slide came down May 2, which prevented further digging. The *Seattle Post Intelligencer* (May 20, 1898) reports four bodies taken from a second slide a few days after the disaster. Possibly reports of numbers of

---

[2]*Chilkoot disaster*: the Sheep Camp Snow Slide, April 3, 1898. "It is recorded that the bodies of sixty-three men and one woman ... were eventually recovered from the scene (old accounts vary, however, in their lists of the unfortunates more or less." Of these, 51 are listed by name, 29 being shipped "outside" and 22 buried in the cemetery at Dyea (Atwood 53-55).

[3]*The dead...*: Hanable, page 56, gives the names as B. B. Ianthorpe and J. Towhere. Versions vary, and it is possible the dead were from different snowslides.

people vary because people were describing different slides. The number of people who actually died in all the snowslides was not recorded.

Summit from camp below the fifth bench. There are several signs that this photograph was taken after the April snowslide and heavy snows. No one is in camp and sleds are empty, the goods (center) are well covered, and an oar (left) stands as a guide should heavy snowfall or an avalanche reoccur. (No. 26, Roppel-DeArmond Col., April, 1898)

The avalanches buried tons of outfits, some to a depth of 25 feet. Many people became disheartened, especially those whose outfits were swept away, and turned back to Port Valdez. Those who persevered and reached the summit found a gradual slope down (what later was named) Klutina Glacier. The size of the load a man took down the six-mile slope was limited only by his ability to hold it back. Incidents arising from downhill slipping were numerous. Mushers could be seen in canyon-like crevasses examining piles of kindling wood that once had been boxes of supplies. Scattered piles of goods gave testimony to a runaway sled.

Many, including Townsend, camped at the foot of this glacier to consolidate their supplies that had been cached at various points behind them. Beginning at the foot of the glacier, tent cities sprang up all along the trail and at the mouth of the lakes and creeks until, between the summit of the Valdez Glacier and the Copper River, there were ten camps. For four miles the men followed the right bank of the river flowing from the glacier along a level valley, about a mile wide, over gravely, glacial moraines to a grassy meadow behind a big rocky bluff.

This first stop was Twelve Mile camp, reported to be twelve miles from the summit, where the valley widened and timber, consisting of spruce and cottonwood, began to grow. Here a city of between 100 to 125 tents sprang up. Many parties whipsawed lumber and built boats of innumerable styles and models just as at Lake Bennett on the route to the Klondike. As on the Yukon River to Dawson, the Klutina River banks became strewn with wrecks. One army officer wrote, "That there were not more casualties from drowning is remarkable" (Glenn 1899, 344).

Men moved from Twelve Mile camp either by packing or for those with boats, by dragging them down the glacier's outlet stream to Three Mile or Saw Mill where another party sledded over the glacier. At Six Mile Camp, Townsend's party constructed its boats. Then, after a cluster of tents and the portable sawmill that a San Jose, California, man packed to the area, came Seven Mile Camp (which was never large and was completely deserted by July 1898). These camps were named for their distance from Twelve Mile Camp.[4]

As Lake Bennett was the beginning of the water route for the Yukoners, Lake Klutina was the beginning for those who went over the Valdez Glacier. One prospector estimated there must have been a thousand boats built between the foot of the glacier and Klutina Lake (Margeson 1899, 116). Four or six people at a time, each paying $5, could cross the milky waters of the lake on the *Manhattan*, a fourteen-foot steam launch belonging to Dr. Ottaway (of Rochester, New York). After the lake, the men descended the rapid Klutina River, which had a fall of some 600 feet in thirty miles. Along that stretch there were more camps: Peninsula City with 125 tents, Amy's Rapids, Cox's Landing, and finally Copper Center at the confluence of the Klutina and Copper rivers.

During 1898, only a few of the hardy hopefuls who crossed the Valdez Glacier ever reached the Klondike by this route; instead, many, like Townsend and his party, remained to prospect in the tributaries of the Copper River and the Wrangell Mountains. They soon learned that gold could not be scooped from the gravel bars by the shovelful, as newspapers in San Francisco and Seattle had implied. Placer gold, because it is heavier than gravel, settles to bedrock, and working deposits found on the tributaries of the Copper River proved uneconomical because of the great depth of glacial gravel.

---

[4]*Twelve Mile Camp*: so called because prospectors believed it was twelve miles from the summit of the glacier. The U.S. Geological Survey geologist assigned by Abercrombie's reconnaissance team in 1898, Frank C. Schrader, measured the distance at ten miles.

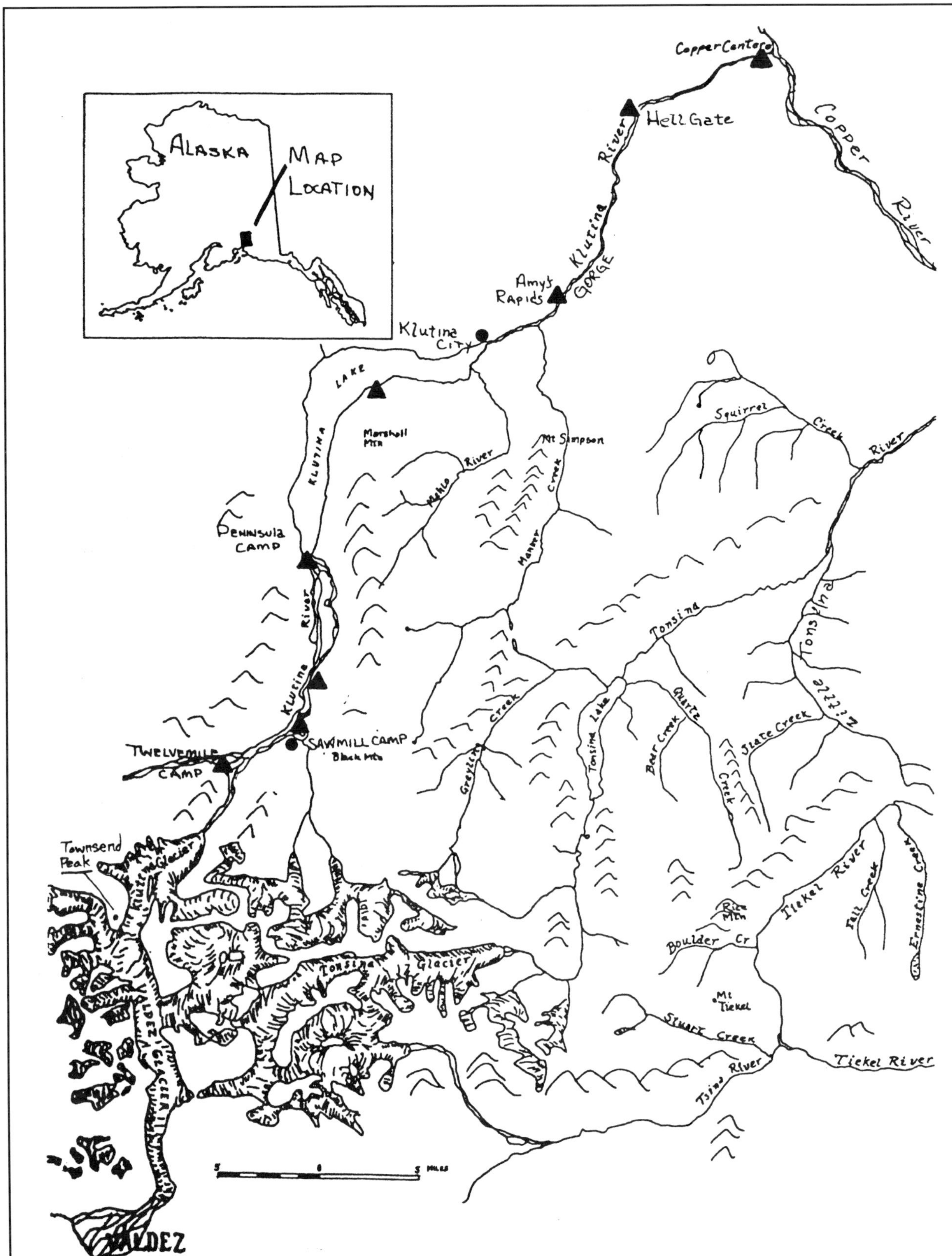

Map of Valdez to Copper Center, after the map of Horace Conger (Holeski 1983, 66).

Some prospectors willingly put back-breaking effort into placer mining, and considerable work was done on Mauker Creek, Quartz Creek (a tributary of the Tonsina), and on Ernest or Earnestine and Fall creeks (tributaries of the Upper Kanata). Shelton Creek, another tributary of the Klutina, was also worked. Enough good gold, coarse as corn kernels, was taken out. Oscar Rohn, a USGS geologist who explored the area in 1898, felt that its presence showed there was gold-bearing rock in the vicinity (Rohn 1900, 402). However, most of the good prospects discovered by the season's work were confined to those of Quartz Creek.

The rush for northern gold brought not only gold seekers but also government exploration expeditions, one of which was in the Copper River country at the same time as Townsend's party. Demands for government assistance in developing a route to the gold fields, preferably all on American soil, caused the Secretary of War to send out three expeditions. The first went overland from Skagway and Dyea to present-day Eagle on the Yukon River. The second went from tide water on Cook Inlet to the Copper River Valley and thence to the Fortymile region. The third penetrated the valley of the Copper River and made its way to the Tanana River. Abercrombie, who had been in the region in 1884, was put in charge of the latter expedition and arrived in Valdez on April 18, 1898.

Where men like Townsend had succeeded, Abercrombie's attempt to man-handle supplies over the Valdez Glacier failed because of the lateness of the season, and Abercrombie returned to Seattle to purchase pack animals. He and his party set out again on August 5 with his pack train and traversed the Valdez Glacier with great difficulty because of soft summer snow, crossed the Copper River, worked his way to Mentasta Pass, and went up the east fork of the Tok River. When he ran out of supplies, Abercrombie and his party returned by raft to Copper Center where they rented a boat from W. S. Amy and headed down river, successfully navigating the rapids to tide water (Abercrombie 1900, 297-351).

The discouraging prospects of finding quantities of gold led many discontented prospectors to leave the country as winter approached. Some returned over the glacier to Valdez, but more followed Abercrombie's example and took boats down the Copper River.

To aid the retreating, destitute prospectors, the government set up relief stations. On August 5, 1898, Charles Brown, a government employee of some forty years, was appointed by the Assistant Secretary of War as quartermaster's agent for the relief of destitute persons at Valdez. The majority of the men who made the return trip over the glacier were in an exhausted physical condition and required several days to recuperate. From August to October 1898, 306 men were carried south gratis by steamers of the Pacific Steam Whaling Company (Glenn 1899, 451). Brown's men also supplied food and fuel to several relief stations. These had been constructed by Abercrombie's parties in the spring of 1898. One station was on the second bench and the other at the foot of the summit. Each consisted of a tent stretched over a skeleton frame and contained an oil stove and necessary cooking equipment. Townsend would later seek shelter in one of these stations.

By November 1898, horror stories of crossing the glacier were being told in Valdez. Although Mike Smith traveled with a party of eight, when he gave out within a mile of the summit, he froze to death. The others were unable to render assistance without endangering their own lives. The thermometer of another group read 48 degrees below zero, and there were three feet of new snow underfoot. The members of another party dug a hole with snowshoes on the summit when a storm broke. They remained there four days, subsisting on four quarts of beef stock and ten beef capsules, but Henry Krohn reached Valdez in such bad condition that he died.

Of those who had sufficient provisions and were determined to winter in the Copper River Valley so they could continue prospecting in the spring, the greater number settled at Quartz Creek or Copper Center. A few were scattered in isolated camps along the Klutina River and along the upper Copper River. The exceptionally cold winter of 1898-1899 was a disaster for these 400 or more miners. Most were unprepared for the rigors of an arctic winter. As lengthening nights and deepening snow crept in on them, many grew uneasy and wondered if it had been a mistake to stay. Then a mysterious sickness, universally diagnosed as black rheumatism, frost bite, and everything but scurvy, struck, and prospectors were driven into a fever of desperation.

"As early as the seventeenth century scurvy had been recognized as being somehow related to a lack of fresh fruits and vegetables in the diet" (Fortuine 1992, 306). Enlightened sea captains took precautions, such as serving lime juice, to prevent it, but only a few Copper River prospectors knew to pack such preventatives. The symptoms of scurvy come on gradually, being recognized by a failure of strength and the onset of exhaustion at slight exertion. The *Chicago Record's Book for Gold Seekers* gives the following lurid description of scurvy:

> The countenance becomes sallow or dusky, eyes sunken, and constant pains are felt in all the muscles. After some weeks utter prostration ensues; the appearance is most haggard; great trouble is experienced with the mouth, sore gums, and teeth falling out; the breath is extremely offensive; finally come swellings and dark spots on the body, with bleeding from the mucous membrane; then painful, extensive and destructive ulcers break out on the limbs; finally diarrhea, pulmonary or kidney trouble may give fatal result. (*Chicago Record* 1897, 236)

The first cases of advanced scurvy were reported in mid-January by Dr. Quick, and he began to look for a cabin in Copper Center to use as a hospital. A week later Dr. Nierman gave up his cabin with no strings attached because he expected to be paid by the government. The two doctors opened the hospital on January 23 with Luther Guiteau as record keeper, helper, and cook. It is Guiteau to whom we are most indebted for the few descriptions we have of how Townsend conducted his hospital service during February, March, and part of April 1899 (Guiteau 1941, 16-18). If Townsend wrote letters during this period, they are not known to exist.

On the last day of January, Dr. Townsend arrived to assist Dr. Nierman because the latter had no previous experience with scurvy and had just about exhausted his resources and patience. Dr. Townsend immediately recognized that all the cases were severe, but he thought he could save the men if the medicine and supplies held out. He arranged for a man named Brandon, who had a big Husky dog team, to go over the Klutina and Valdez glaciers to seek help. Brandon successfully crossed the glacier, bringing the following appeal from Townsend to Quartermaster Brown, dated February 3, 1899:

Dear Sir:

I beg to report to you herewith the serious condition which prevails in this camp. Scurvy has developed to an alarming extent. Two deaths have occurred and the hospital, which it was necessary to establish, is now full, and still throughout the camp are many who should be admitted. Up the Copper River the condition, I understand, is equally serious. Many are wholly without means and dependent upon others for nursing, medical attention, etc. We are in need, too, of some medicines, and any fresh fruit or vegetables which may be had at Valdez. Two have volunteered their services to bring in these supplies and are now on their way to Valdez. Several men are suffering with frozen feet, and amputations in a number of cases will, no doubt, be necessary. If it be within your power to give any assistance, I would respectfully ask your immediate consideration. (Abercrombie 1900, 38-39—and in context, Appendix A)

Brown, according to his official report, was barely able to respond to Dr. Townsend's request (Abercrombie 1900, 37). The most he could furnish were some green potatoes and sauerkraut, and these meager supplies were taken into the interior by volunteers who succeeded only after being foiled in three attempts to cross the storm-ridden glacier.

There were twenty patients in the makeshift Copper Center hospital that first week of February, according to Guiteau, and most responded well to Dr. Townsend's treatments (Guiteau 1941, 16-18). Two men stopped at the hospital on their way out of the Copper River Valley and donated their medicine,[5] which helped. All along the Copper River Valley, as men's condition worsened, many moved to Copper Center. A few, because of badly frozen feet as well as debilitation from scurvy, had to be hauled in. Many men were in such poor physical condition due to scurvy that their feet could not be amputated until symptoms abated as hemorrhaging could result (see Appendix B). During those winter months, Dr. Townsend found it necessary to amputate some toes and feet.

Until mid-March, the situation remained unchanged. The only able-bodied men were the handful on the hospital staff. The other miners in camp were too ill and weak from scurvy to give assistance, although they had not reached the stage of becoming hospital cases. When W. J. Woliben of Lincoln, Nebraska, volunteered his services, Dr. Townsend gratefully accepted them. He and his staff were nearly worn out from constant long hours and heavy work. During those couple of months, Dr. Townsend, with his scientific curiosity, asked every man who visited the hospital what he or she thought caused scurvy and what aggravated the condition most after it was contracted. Many thought the condition was poor man's gout from eating too much salt meat and aggravated by stale "grub" and not enough exercise (Remington 1939, 87-88).

From February through April 1899, most of the men who were well enough, and those whose illnesses still allowed them to get around, decided to hazard the glacier with its known terrors rather than stay in Copper Center and succumb to the horrors of an illness they did not

[5]*their medicine*: Guiteau did not describe what kind, but Townsend in his scurvy report, lists some, including potassium chlorate, boric acid, borax, carbolic acid, mild laxatives, and anodynes for pain, resorting to morphine if necessary (Appendix B, 4 paragraphs from the end). (AP)

understand. Brown's report chillingly details the dire extremes encountered by parties who returned over the glacier from February to April 1899 (Appendix A). Men traveled in large parties as a means of surviving the terrible storms sweeping the Valdez Glacier that winter. Despite the hope of safety in numbers, the men often had to be rescued by Brown's relief parties. From February to April 9, Brown's rescue teams, often equipped with dog teams and supplies, helped bring many people off the glacier. Often relief was too late, and the men returned with bodies, such as that of Pete Benson; he died after falling into a creek and freezing to death before he could be removed to a place of shelter.

A coroner's inquest of another tragedy provides more details. On the last Friday of February, six men left Twelve-Mile camp. Two had scurvy, Adolf Ehrhardt and Maximillian Miller, and these had contracted for $150 with the other four to bring them on sleds down the Valdez Glacier to port. Both scurvy victims were partners in the Scientific Mining Company, and those bringing them out were Rudolph Ellerkamp, a correspondent for the *Louisville Post*; Alfred Elliman, another Scientific Mining Company partner; August Schultz, president or vice president of the Tanana Mining Company; and a Dr. Logan, who headed the party. Axel Elder passed them on his way down the glacier and arrived at the relief station that evening.

A freezing gale brought the party to a stop at the foot of the fourth bench, about fourteen miles from Valdez. J. B. Ridge said he saw "the party on the glacier at a distance. The storm was raging so I could not render them any assistance." When the group did not turn up at the relief station, five men set out to find them. Three of the missing party were found together, frozen. The men with scurvy, Ehrhardt and Miller, were on their sleds, rolled in blankets, and Schultz was "in a snow hole," which he had apparently dug for refuge. All three had frozen to death. The dead men on sleds were taken down the glacier that night, and the next day a search for the other three resumed. About a mile further down the glacier, two more bodies were found in a twelve-foot crevasse, side by side, frozen to death. Dr. Logan, mentioned in Townsend's letters, was never found.

Abercrombie returned to Valdez in spring 1899 because he had received orders to open a military road from Valdez to Copper Center and from there to proceed by the most practical route to Eagle City on the Yukon River. He and his party disembarked at Valdez on the morning of April 22, 1899. His impression as he arrived was

> That they [the prospectors] had passed a terrible winter, beyond all question of doubt; that many of them had died from scurvy and being frozen to death was in evidence at the little graveyard that had sprung up since my departure the year before. (Abercrombie 1900, 15)

Because of the inhabitants' long hair and beards, Abercrombie had difficulty recognizing many of the men he had met the previous year. They were a "motley looking crowd," he said, dressed in "mackinaw suits of all varieties and colors, faded and worn by exposure to the elements and their long journey over Valdez Glacier" (Abercrombie 1900, 19).

Townsend's letters resume in the spring, and he describes his return trip. Upon his arrival in Valdez, he reported to Abercrombie that five sick men were at Barrett's camp at the foot of the Klutina Glacier, and Abercrombie authorized Brown to send nine men and two teams of four dogs each to transport the sick men. The rescuers arrived at 7:00 p.m. on the same day. On the next day, April 28, they were unable to start due to a storm. The glacier at this time of the year was awful to witness, being composed of glaring ice and huge almost perpendicular benches extending to a great height (Brown, Appendix A). It required four or five men to assist a sick person on a sled down the steep benches and to cross strips of ice so narrow that a traveler's misstep would drop him into crevasses hundreds of feet deep. The rescue, however, was accomplished without mishap.

Brown, on Abercrombie's orders, rented a house for a hospital in Valdez. However, Abercrombie had brought in a government doctor, so Townsend's services were no longer needed. A note of dejection over his few prospects at the end of a great adventure comes through in the final letters Townsend wrote home. But Townsend was called upon to respond to one final service.

At the request of Abercrombie, Townsend wrote in "arresting clinical detail about the natural history" of scurvy, detailing some of the things he had learned from questioning the patients in his hospital. Abercrombie included Dr. Townsend's report with his own and other reports to the War Department (Appendix A). Dr. Townsend's prescription for a cure still makes excellent sense (Fortuine 1992, 307).

Too late for Townsend and his patients, Abercrombie's party set up relief stations at the mouth of Grayling Creek, at Amy's Rapids, and at Copper Center and detailed enlisted men "to extend relief and encouragement to the demoralized and destitute prospectors" (Abercrombie 1900, 19). The stations stayed open until the latter part of September 1899, long after Townsend had returned home. Just prior to withdrawal of the personnel at the stations, the last of the scurvy patients, a man and a woman living at a miner's camp near the head of Quartz Creek, were packed out on horses to Valdez. Abercrombie reported that the army provided aid in Valdez and the Copper River Valley to 480 persons, representing all nationalities, both black and white, and all professions (Abercrombie 1900, 19).

Yet Charles Remington, a prospector, depicted the army men as less than generous. He wrote that those in charge of the relief stations refused rations to starving prospectors, claiming they had orders to hang on to the provisions. He also reported a trailside encounter with an army party that was carrying fresh provisions; the army men boldly requested some sheep meat from the scorbutic prospectors in Remington's party without offering any fresh food to them in return (Remington 1939, 137, 149-50). This may have been an isolated incident as no other such complaint has been found.

In his report, Abercrombie offered an analysis of what took place at Valdez and the Copper River Valley during the stampede of 1898-99. He blamed ignorance for most of the trouble. He concluded that the men who came to Valdez seeking a short route to the interior had no idea what to expect. Three-quarters of all outfits in 1898 had turned back, he wrote (Abercrombie 1900, 19). Those who did not were to decide later that they were foolish not to have turned back. Abercrombie categorized general causes for failure among the stampeders to survive the rigors of the north: First was the advanced age of the prospectors, "which averaged over 47 years." Second was the lack of mining experience:

> It is a well-known fact to the professional miner that his calling involves not only theoretical knowledge of the structure of the earth's crust to detect the impregnations of a mineral-bearing zone, but also a wide experience in the economic development of the same after discovery. Of the 4,000 persons who composed the stampeded to the Copper River Valley in 1898, I do not think a man of such caliber was to be found in the ranks. Third was no business qualities.

Abercrombie, who tended to exaggerate, wrote that ninety-five percent of the men were already business failures who hoped to make good in the gold fields (Abercrombie 1900, 20).

Some of Abercrombie's statements are colored by quick conclusions and sometimes factual errors. Brown estimated the average age at 40 and qualified it by adding that many men of old age were among the number, which tended to increase the mortality rate (Glenn 1899, 453). Abercrombie also wrote that two-thirds of the people attempting to cross the Valdez Glacier were frozen to death (Abercrombie 1900, 19), but Townsend's letters, other memoirs, and Brown's report do not substantiate such high figures.

By the spring of 1899 most of the sick miners had either recovered or been shipped out of Alaska. One of Abercrombie's men, John F. Rice, crossed the glacier in late May 1899 and found 10 or 12 cabins deserted except one at the foot of the Klutina Glacier. At Amy's Landing, he found 50 cabins with only one occupied. Rice established a station at the rapids and left Private Garrett in charge of the military stores, and Garrett "incidentally operated" a ferry across the Klutina River. At Copper Center, Rice estimated 60 inhabitants, 30 of them Natives catching their winter's supply of fish (Abercrombie 1900, 96-97).

In the years to come, where Townsend and his party prospected among the Tazlina, Tonsina, and Tiekel Rivers, negligible placer gold and only small amounts of gold from hard rock mines were found (Cobb 1973, 23-32; Berg 1967, 48-52). Had Townsend gone farther up the Copper River near the head of the Chistochina and Slana Rivers, he might have had better luck. An estimated 150,000 ounces of placer gold were produced over the ensuing years (Cobb 1973, 26).

It was the mineral for which the Copper River was named that made the area famous. During the summer of 1899, some fifteen or twenty persistent prospectors found heavily mineralized zones of copper, bornite, and other ores. Abercrombie was mistaken in his belief that no men had sense enough to develop the mineral resources. These copper discoveries eventually led to the development of the famous Bonanza and Kennecott copper mines, which until 1938 produced tens of millions of dollars worth of copper.

As for the Valdez Glacier trail, after the fall of 1899 it was only infrequently used when the Trans-Alaska Military Trail through Keystone Canyon had been forged for ninety-three miles into the interior (Lethcoe 1996, 110). For the later gold seekers who came north for gold, the Copper River district no longer held much interest, nor did the Klondike for that matter. In 1898, gold had been found on the beaches in Nome, and by the summer of 1899, the great migration of gold seekers had turned to the golden shores of the Bering Sea.

THE LETTERS

Envelope of letter number 1, postmarked SEATTLE, WASH, FEB 3, 1 P M [18]98.

Letter 1 Postmarked SEATTLE, WASH, FEB 3, 1 PM, 98

Seattle, Washington
Feb. 3, 98

We arrived safely in Seattle yesterday at 11:30 a.m. Our journey was an uneventful one. We were no way interfered with by storms, which we were somewhat fearful of on the Great Northern [Rail]road. Indeed this has been a most remarkable season in the West. In the memory of the oldest inhabitant it has never been so mild. In the Dakotas and in Montana there has been hardly any snow, and it was not until we reached the Rockies that we found any great fall. In the Cascades it was deep, though, and in some places we passed through drifts as high as the car windows. Coming out of the mountains on this side, the change is a remarkable one. Within a few hours from winter cold and snow and ice, the air becomes balmy and spring-like. You hardly require your overcoat [any] longer, and the feel of the atmosphere makes you think of spring at home. Our Mr. Egan[1] had gotten us quarters, and we immediately took up our life at # 1711 2nd Ave. We will now proceed with our outfitting. There will be no difficulty in obtaining all supplies here and at reasonable prices. All you see in the windows is Klondike outfittings. On the street, for advertising purposes, are Klondike sleds drawn by dogs of all sizes and colors. Various other Klondike machinery is being exhibited here and there.

Klondikers are everywhere and are arriving at the rate of 1 or 2 hundred daily if reports here are true, and I have no reason to question; for on the train with us there were no less than 100. Most of these, too, strange to say, were bound for Copper River. We are going to have lots of company. The reports are all favorable. Yesterday, though, a miner returned from Port Valdes[2] and reported the pass still closed. He also gave news of a double murder and a neck-tie party following. The murdered men were from Salem, Ohio; the murderer, a Montana cowboy.[3] The amount of gold which will come out this spring will be immense according to the opinions of those with whom I talk, and everyone seems to think it a good thing.

As I write here in the Diller Hotel, there are no less than ten around me talking "Klondike." Mr. Creighton was down yesterday and inspected the boat on which we sail and reports her as being in excellent condition and that we have the best of staterooms. She will probably not sail,

[1]*Our Mr. Egan*: His affiliation with the Keystone Company (KC) is unknown. (AP)

[2]*Port Valdes*: Valdez is the modern spelling, from 1899 on. "Valdes" is preserved except as LST himself makes the change. (AP) *Valdez*: Valdez is located at the head of the estuary of the same name in Prince William Sound. After the 1964 Good Friday earthquake, which, along with a devastating tidal action, severely damaged the town, it was relocated on firmer ground a few miles to the northwest. (PR)

[3]*The murderer, a Montana Cowboy*: The murders and hanging occurred January 2, 1898, at the beach at Port Valdez where a few crude log cabins stood, remnants of the Indians' former trading excursions over Valdez Glacier. Among the early arrivals that winter was a party of Eastern prospectors who took up residence there. In Seattle, they outfitted a cowboy, M. F. Tanner, known as "Montana Cowboy," hoping to use his wilderness experiences. He proved to be arrogant, and one night the partners discussed cutting him out of the partnership. He overheard them, became angry, rushed into the tent, and shot to death two of his partners, N. A. Call of Worthington, Minn. and W. E. Lee of Salem, Mass. Two others escaped. Tanner was caught and judged guilty at a miner's meeting (a legal official body in mining country where there was no other law). Twenty-nine voted for immediate execution while nine were in favor of conveying him to Sitka. He was promptly hanged. (PR)

however, before the 15th or 20th, a week later than we anticipated. Every day makes me more pleased with Mr. Creighton. We have made no mistake in going under his leadership.

The party are all well and happy.

With love to all,
Leroy

Letter 2 Postmarked SEATTLE, WASH, FEB 11, 11:30 AM, 98

Seattle, Washington
Feb. 10, 1898

Dear Father, Mother, and Sister:–

Letters of the 1st and 2d inst.[4] to hand, that of the 2d having been received this a.m. No doubt you have before this received mine under date of the 3d. Everything is moving along smoothly. Today we are having our groceries and our hardware packed, and we expect to have things in good shape by Saturday, though we do not get off until Tuesday. I have an excellent personal outfit and feel that I shall not have forgotten anything, having compared several printed lists in making the selection. Our party now numbers twelve; and unless Mr. Barnum joins us, I think that is the number which will sail. Mr. Sharp, one of the Pittsburg men, has dropped out but has taken as substitute a Mr. Clark of this city, who has had experience in mining and with whom I am much pleased. Mr. Clark has been known to Mr. Creighton for some time. We have also gotten a professional cook to look after the commissary department. He is quite an acquisition, I assure you, and will save an immense amount of worry and trouble, besides giving us better prepared food. He is strong and of a good disposition and presented excellent recommendations.

We have an excellent party; and if all goes as satisfactorily as it has, we shall have no dissension whatever.

Boats are leaving every day for the diggings with every berth sold. With regard to price of tank and pump offered Dr. Grim [in Beaver Falls], I believe $15 was the amount. I cannot now be sure, however. Let him have it, though, if that is his recollection of the amt. This includes tank, pump, and the connections only.

Tell C.W. that I received his letter with that of Little Edith[5] enclosed. Package by express also rec'd, and Mr. Richards expresses thanks. I will write them shortly.

I will write again before leaving.

With love to all,
Leroy

---

[4]*inst.*: "instant," said of the current calendar month, as in "the 10th instant," i.e., the tenth day of the current month; often abbreviated "inst." (AP)

[5]*C.W. ... Little Edith*: C.W. was Charles Walker Townsend, Leroy's uncle. Edith, C. W.'s daughter, was born August 4, 1894; she was 3½ years old at the time. (PJT)

Letter 3[6] Postmarked SEATTLE, WASH, FEB 16, 10:30 AM, 98

Seattle **2/15 98**
7:40 p.m.

Dear Father, Mother & Sister:–

We have taken everything aboard and expect to sail within a few hours. Everything O.K. Got your letter this p.m. Am glad all are well and pleased to hear that friends are interested as you write. Things have gone smoothly, and all are in the best of spirits. We are just settling up with the Co. for freight, and I am just taking a few minutes to write. You will next hear from me at Port Valdes. Address all mail c/o F. H. Whitworth, #417 Seneca St., Seattle, Washington. He will forward same. The *Alliance*[7] is full. She is small but a good staunch vessel. I am sending Al[8] the spools of exposures made on the trip. See him.

With love to all,
Leroy

The steamer *Alliance* on which the men of the Keystone Company sailed from Seattle on the morning of February 16, 1898, to Valdez. (Snowden Col., Townsend, February 1898)

[6] *Letter 3*: This letter is written on the back of a pre-printed form for E. E. Caine and Co. of Seattle. (PJT)

[7] *Alliance*: Built in 1887 in Astoria, Oregon, as the *Alliance* of the Portland & Coast Steamship Company, she was later almost destroyed by fire, and the remains were sold to California parties who had formed the Copper River Development Company. Rebuilt, she operated between Puget Sound and CR points. The Pacific Steam Whaling Company added the *Alliance* to its fleet about the time that Townsend went north. In March 1898 she suffered the first of several mechanical problems and in January 1900, by then renamed the *Townsend*, she suffered another such failure and stranded near Haines, Alaska, becoming a total loss. (PR)

[8] *Al*: J. A. Snyder, friend of LST to whom film was sent for developing—also referred to as J.A. (AP)

Map of Inside Passage route to Juneau and Skagway. From Juneau, the *Alliance* would proceed NW into the Gulf of Alaska.

Letter 4 Postmarked MARY ISLAND, ALASKA, FEB 19, 1898

In Greenville[9] Channel,
Saturday, 9:50 a.m.
Feb. 19, 1898

Just a few lines which I will post at Mary Island[10] where we will stop sometime this p.m. We steamed out of Seattle Wednesday morning at 8:30, being later than was anticipated when I wrote Tuesday. The morning was ideal; light fleecy clouds scudded overhead, and here and there a great cumulus or summer day cloud shone silvery in the sun. The mountain peaks, white with snow, looked on in silent wonder. We sailed smoothly along, enjoying the ever-changing scenery until about noon when we struck rough water in crossing Juan De Fuca Strait.[11] This greatly changed the scene; and soon the rails were filled with passengers feeding the gulls, who as some poor unfortunate gave up his meal, would swoop down and fight for the spoils. The sea is no respecter of persons, and I gave to them my share with as good grace as possible.

Three o'clock found us in calm water again and enjoying ourselves once more. Thursday brought rain and snow and a cold wind, but no rough seas. Yesterday, Friday, we were not so fortunate, however; for with the dawn came snow and hail; and as we crossed the open seas at the head of Vancouver Island, the water was very choppy; and again we witnessed the return of the swallows. Only two of our party had any appetite for dinner. About three o'clock we again reached the quiet waters and since then have been enjoying ourselves again. We are now about

---

[9]*Greenville*: Spelled *Grenville*, this channel is in British Columbia, south of Mary Island and east of Pitt Island (see map previous page), named by Vancouver in 1793 after Wyndham Grenville, a son of George Grenville of Buckinghamshire (Walbran)

[10]*Mary Island* is between Ketchikan to the north and the Canadian boundary to the south in Revillagigedo Channel (north of Duke Island). Mary Island became the port of entry for Alaska in 1892. All vessels traveling from Canadian waters into Alaska had to pass through customs there. In 1900, the customs house was transferred to Ketchikan. (PR)

[11]*Juan De Fuca Strait*: South of Victoria in B.C. and north of Port Townsend in the U.S., named after a Greek pilot in the service of Spain who claimed to have found this waterway in 1592 (Walbran).

the head of Greenville Channel and will shortly pass the wreck of the *Corona*,[12] which left Seattle since our arrival there. You have probably read of her loss in the papers.

**** While writing the above the *Corona* wreck was sighted, and I went on deck to see her. She lies half submerged, her bow shoved well in the air and pointing toward a small island on the left of the channel through which we were passing. She had about 250 souls on board; and with the exception of two stowaways, all were saved. The stowaways were in the hold and were lost. She sank within about 60 - 75 feet of the shore.

The *Corona*, run aground. (Puget Sound Maritime Historical Society, # 1567, February 1898)

[12] *wreck of the Corona*: On January 23, 1898, the steamer *Corona* ran into a reef between Porcher and Kennedy islands near the mouth of the Skeena River. All 328 passengers on board the Pacific Steamship Company vessel escaped to a nearby island in lifeboats, but two stowaways in the hold drowned. The 220-foot ship, built in Philadelphia in 1888, sustained $30,000 in damage, but she was patched, pumped out, and refloated on March 23, 1898. She was towed to San Francisco in September of that year. The pilot was cleared of blame for the accident. At the spot she struck, the charts showed a depth of seven fathoms with no indication of rocks. This gas schooner gave service until August 26, 1929, when she caught fire off Noyes Island and foundered. (PR)

After passing the wreck, we shortly ran into rough water; and at noon I had no appetite for dinner – am writing at 12:45. At 12:25 we ran to the lee of a small island and anchored, where we are now rolling merrily on the bounding billows. Without, the wind is blowing lively, and it is spitting snow, probably the beginning of a blizzard. Everybody is within doors, keeping warm. In Social Hall, which is next my berth, they are playing cards to a merry tune played on the Regina box.[13] On account of anchoring here, we will be late in reaching Mary Island by the p.m. We may not make that point before tomorrow.

6:20 p.m. Since our coming to anchor, the storm has let up, but it is still so hazy that anchor will not be weighed tonight. The tide is out, too, and all around us the water is breaking over the reefs. The ship's boat has just been manned and a party have gone in her to an Indian village[14] about two miles up the bay. We will make Mary's Island tomorrow, and this will be mailed by the captain there. All our party are well and happy. I hope all at home are in good health and not fretting about me.

With love to all,
Leroy

P.S. Remember me to friends.

Letter 5 [No envelope]

S.S. *Alliance,* Monday
Feb. 21, [98] 11:45 a.m.
off Prince of Wales Island,
Due at Ft. Wrangel 4 p.m.

Dear Father, Mother, and Sister:–

I wrote you yesterday and mailed letter at Mary Island, which point we made about 4 p.m. There is a government inspector at that place, and the ship's waybills had to be examined. We were there about half an hour. The day had been most disagreeable: cold, windy, and spitting snow, indeed a regular snow storm. About 4:20 it cleared off, though, and we weighed anchor and proceeded on our way. This was unfortunate, for we had not traveled long until a regular blizzard struck us. It was impossible to see 50 ft. ahead, and the wind was fairly a gale. We concluded to make an anchorage and were proceeding slowly with this in view, when the lookout called out "Reef to starboard," and a moment later, "Island ahead." The pilot put the vessel hard to port, and we missed the reef. The island we could not escape, however; and a moment later, with a grating sound, we were aground. This occurred at 5:50. I was in our stateroom with Mr. Rayl; and as I heard the rudder ropes grating through the guards, followed by the signal to back,

[13]*Regina box*: This is the Regina music box popular in that day. The music was recorded on large, flat metal disks, which could be changed like a phonograph record. The tone was deep and resonant, nothing like the tinkling little music boxes of today. The disks contained hymns, popular tunes, operetta scores, and other choices. An automatic, self-changing disk player with a slot for pennies came into use in public places around 1896. (PJT)

[14]*Indian village*: Probably this is the Tsimshian village of Port Simpson, near present-day Prince Rupert, British Columbia. (PR)

I opened the door and looked out. As I did so, I heard someone exclaim, "We're going to strike." At the same instant the outlines of the shore appeared, and with a slight shock we struck.

There was little excitement. A few of the steerage passengers were panicky, but these were soon quieted. A few did foolish things. The bow was shoved clear up on the rocks, and we were fearful that the wind, which was terrible, would beat us round and thus completely wreck the vessel. To prevent this a boat was lowered and an anchor placed aboard. Twice they attempted to carry it round the stern to windward and failed. The second time the boat seemed as though it would swamp; and the 2d mate, who was in charge, lost his nerve and called out to be pulled aboard ship. He wouldn't make another attempt. The first mate now took his place, and with four good oarsmen and a, "We'll do the best we kin, Captain," started. Except for the dim spot of light which their lantern gave, nothing was to be seen of the boat. But slowly that speck of light went astern and then to windward, and finally a faint cry told us that the anchor had been cast, and for the time we were safe. As the cry came back, I called for three cheers for the 1st mate and his boys, and we gave them with a will. They deserved them, for it was dangerous work with wind and snow and water to fight. Fortunately, we grounded at low tide; and as our stern was anchored off the shore, we had only to wait for the incoming waters to raise us off. This, with the aid of our propeller, they did at 9:45, making about 4 hours that we were fast. We felt our way out and steaming ahead a short distance came to anchor safely just off Ketchikan, which point we were trying to make when we went aground.

We suffered no injury to the vessel and today are bowling along with bright and pleasant weather and calm waters. All morning the porpoises have been playing about the boat, and water fowl are on every hand. The eagle, bird of freedom (he has plenty of it here), is in evidence. I have seen a dozen or more this a.m. – great bald-headed fellows. Yesterday we passed the wreck of the *Corona,* but I believe I spoke of her before. We learned at Mary's Island, too, of the truth of the loss of the *Clara Nevada* with all hands.[15] What is the Leigh party[16] doing?

With love to all,

Leroy

[15]*Clara Nevada*: Formerly the U.S. Coast and Geodetic Survey steamer *Hassler*, the newly renamed *Clara Nevada* on her first voyage apparently caught fire. She was in Lynn Canal returning from Skagway and Dyea to Seattle on the night of February 5, 1898. A gale of near hurricane force was blowing at the time. The sunken vessel was located a week later on Eldred Rock with only the spars above water at low tide. There were no survivors. No passenger list was maintained, so the exact number aboard is unknown, but it was estimated to be over a hundred. (PR)

[16]*Leigh party*: This is another prospecting concern, the Beaver Valley Mining and Prospecting Company of which William H. Leigh was the treasurer. It consisted of about ten members, most of whom were from Beaver Falls and New Brighton, Pennsylvania. They apparently left for the Yukon River in Alaska about April 1, 1898. Leigh sent word from British Columbia in July that he was returning to Beaver Falls. His success or lack of it was not reported in the Beaver Falls *Weekly Tribune* account of July 26, 1899. (PJT)

Above and below, passengers bound from Seattle to Valdez, February, 1898. At least one Keystone Company member's face is prominent, above left, with beard—believed to be Rev. Orphanus Richards. **Below**: passengers are in stances that signify a cold day; a dog, lower left, sniffs the entrance to the forward hold. (Snowden Col., Townsend, February 1898)

Letter 6 Postmarked FORT WRANGLE,[17] ALASKA, FEB 24, 98

Juneau, Alaska,
Feb. 23, 1898

Dear Father, Mother, and Sister:–

Reached here safely at 4 a.m. Am writing at 6:30 in store. Leave at 7:00. Weather cold and crisp. Juneau quite a place. Have had no more mishaps. Yesterday was pleasant, though we ran into several flurries of snow. Last night it was rough. Have sent Snyder a roll of film. They will probably be of little value on account of the conditions under which they were exposed.

Hope all are well. The party are in good health.

With love,
Leroy

Letter 7 Postmarked ORCA, ALASKA, FEB 27, 1897[18]

Aboard S.S. *Alliance*
12:20 p.m. Thursday
Feb. 25, 1898

The day is a beautiful one, the sun shining brightly and hardly a cloud in the sky. We are bowling along at a 10 knot speed through waters that are not disturbed by a ripple. Though we are hemmed in on all sides by snow-capped mountains, it is pleasant on deck; and even an overcoat is hardly a necessity. Last night we had a rough time of it, though. The boat rolled and pitched so that it was almost impossible to stick in one's bunk. Several, indeed, were thrown out, many seasick. I have gotten my sea-legs though now and did not suffer. We should reach Orca[19] in about an hour. 15 - 20 passengers will get off there. The nearer we come to our destination, the more discouraging are the reports as to our ability to cross Valdes Pass. Port Valdes we shall reach tonight. We are not the least discouraged, however, as to our chances of getting over the divide. Quite a number, though, have found their supply of nerve insufficient and will come back with the boat, I understand.

It seems strange to think that as I write to you just after dinner, you at home are thinking of your supper; and I have no doubt you will wonder where I am and what about. I shall write you again at Valdes, which letter and this you will quite likely get at the same time. Send me any clippings from B.F. [Beaver Falls] papers that may be of interest and remember me to friends, Al, C.C., Fickinger, Tatro, etc.

With love to all
Leroy

---

[17] *FORT WRANGLE*: Steamers stopped at Fort Wrangel, today's Wrangell, to unload freight for the passengers who were taking a different route to the Klondike. Gold seekers went up the Stikine River to Telegraph Creek, thence to Lake Teslin using an old Hudson Bay Company trail. From the lake, the miners were within eight or ten days of the Klondike, with smooth water and no portages. (PR)

[18] *1897*: The year is obviously a year off. On the back of the envelope, there is a "Seattle, Wash. Transit" postmark of March 7, 1898. February 25 is a Friday, not Thursday, as Dr. Townsend has written it. (PJT)

[19] *Orca* was the site of a salmon cannery owned by the Pacific Steam Whaling Company, a San Francisco-based firm. Orca was located on the southeast shore of Orca Inlet, two and a half miles northeast of present-day Cordova. The cannery was constructed on that site in 1895. The firm had a cannery in Prince William Sound before that time. (PR)

Leroy Townsend (left) and J. A. Snyder (right). Townsend sent Snyder rolls of film taken on the expedition. This studio photograph is probably taken near Beaver Falls, PA, between 1894 and 1897.

Letter 8 [No envelope]

S.S. *Alliance*
In Prince William Sound
Friday eve. 9:30 p.m.
Feb. 25, 1898

This afternoon I wrote you and mailed letter at Orca, at which place we arrived safely at 4 o'clock. In it I said that the reports regarding the possibility of getting over the pass were discouraging. At Orca, however, I met Dr. Allen, superintendent of the Pacific Whaling Co.'s canning there, who tells us that the pass is now open and that quite a number have already crossed. We have been able to locate the party led by Dr. Winslow, too, also that of the Remmick Co. This all tends to confirm that which the Col. [Creighton] told us in B.F. We shall reach Port Valdes to-night; and to-morrow, after learning what I can there, I shall add to this. To-night we are turning in early in order that we may be prepared for a hard day's work. We got very little sleep last night on account of its being so rough. This is a beautiful moonlight night with the thermometer standing about 28°.

10:00 p.m. Saturday,

We arrived at Valdes Bay last night and anchored. This morning we looked for a landing and finally weighed anchor about noon, went to the head of the bay and attached ourselves to the ice; and since then we have been working like turks getting our goods off and packing them to shore about a mile off. We have not yet finished and will work in turns all night. I am fortunate and will not be called until 6 o'clock. There are about 140 men and 2 or 3 women who have wintered here. They are just now commencing to cross the glacier, and the reports are favorable. All are well. I have taken several pictures of the landing and of Valdes Glacier and pass. I hope they will be good.[20] Will send the spool to Snyder by return boat if possible, together with one exposed in coming up.

I hope I may hear from you before leaving Valdes, but I presume I will be disappointed, for there will be no chance to notify at Seattle for forwarding. We may be here a week or so. If you do not hear from me now for 3 - 4 months, do not be uneasy. It will probably be most difficult in getting mail out of the interior.

I promised Mr. Leigh to write him but will not have time. Tell him about what I have written. Tell J.A. that I will write at length when settled, and ask him to be patient.

With hope for the future and love to all,
Leroy

[20]*... several pictures ... hope they will be good*: Some of these photographs did not turn out, as did not, apparently, many others that he asks about, or they were lost with correspondence in transit. (AP)

Probably boats catching a tow from the *Alliance* into Port Valdez. Charles Margeson in *Experiences of Gold Hunters in Alaska* reports becalmed conditions, adverse winds, and tides made a "pull" helpful. Sometimes steamers demanded a charge of up to $150, as in Margeson's case (27); others provided the service gratis. (Snowden Col., Townsend, February 1898)

Opposite page: Landing supplies on the ice in Port Valdez, Alaska, February 26, 1898. It is possible to imagine Townsend taking a break from his share of unloading activity and walking forty paces from the ship to capture the event with his camera. (Snowden Col., Townsend)

Letter 9 Postmarked SEATTLE, WASH, April 13, 11 AM, 98.[21]

Aboard *Alliance*
Valdes, No. 1
March 4, 98

Dear Father, Mother & Sister:–

This afternoon a number of us have spent on board the *Alliance,* which came in last night. We wanted to see the Capt. and crew. Have had a pleasant time. This morning we went for wood and got back about 12 p.m.; and as we expected to work over the benches tonight, we devoted the p.m. to pleasure. We have the greater portion of our outfit at 10 mile camp; and if good weather prevails, will get to timber beyond Camp only by Monday or Tuesday next, then pack down [the other side]. Have refused $750 cash for our shanty and 3 lots. The company owning the *Alliance* talk some of putting in a wharf on our waterfront. This looks rather favorable, doesn't it? If gold is found on C.R. [Copper River], we've made a hit. Have rec'd no letters yet, may not get [any] for weeks, perhaps months. Wrote you at some length last Wednesday[22] – all well. Say nothing about above. Lots of snow. Leaving for camp 6 miles away – 8 p.m.

Love to all,
Leroy

Letter 10 Postmarked SEATTLE, WASH, MAR 17, 8 PM, 98[23]

In Camp, Copper City [Port Valdez],[24] Alaska
Sunday, March 6, 98, 7 p.m.

Dear Father, Mother, and Sister:–

The *Excelsior* [25]and the *Alice Blanchard*[26] are both due, and I shall have the opportunity to mail this upon their arrival. We have had splendid weather since reaching here and are now comfortably settled. Our cabin is finished, and we have been fortunate in being able to locate a hundred & sixty acres, which, if the route across the Valdes glacier proves good, will be valuable. It is partly in timber, just at the head of the bay, and about 5 miles from the foot of the glacier.

---

[21]*Letter 9*: On the back of the envelope, "Beaver Falls Rec'd, Apr 18, 1898." Also stamped on the back is *STR. ALLIANCE*, large and in purple ink. This letter is written on a postcard, "Postal Card - Once Cent," but written on both sides, folded, and sent in the envelope. (PJT)

[22]*Wrote at length...*: There is no extant letter for Wednesday, March 2. On Wednesday, February 23, he did write, but hardly at length. The letter to which he refers, together with some later ones, apparently became lost in transit. (PJT)

[23]*Letter 10*: On back postmarked "Beaver Falls Rec'd Mar 23, 1898." An extra 2¢ stamp is attached to the printed, pre-stamped 2¢ envelope. (PJT)

[24]*Copper City*: Port Valdez in the early days was sometimes referred to as Copper City. (PR)

[25]*Excelsior*: See note next page.

[26]*Alice Blanchard*: This former halibut schooner plied Alaska waters until being driven ashore at West Bay in March 1929. (*Shipwrecks*, AP)

It is simply astonishing how one is deceived in distance here. What appears a couple hundred yards is a mile. From our camp we look toward the glacier, and it appears only a rifle shot away, yet it is a good 6 miles. It has not yet ceased to be a wonder to me.

S.S. *Excelsior*[27] probably in Seattle. (Puget Sound Maritime Historical Society, # 1694, circa 1898-1900)

I hope the pictures sent Al will be good; they will be most interesting and of no little value. I shall send with this letter a spool showing our camp on our ground in Copper City, our cabin, etc. and showing at the same time the general camp of the seekers after gold on this side of the pass. I have had to make many exposures under conditions that cannot give good results; but if I get even an outline, it will be to me a satisfaction.

When we landed here, it was upon the ice and about a mile from where we are camped. I tell you, it was hard work to pack our goods in to land. The ice was smooth and rough by turns, and we had no ice creepers.[28] The Col. thought we would have no use for them. They are useful, though, and should be made of the best steel and strong. Several of the boys, not our party, were spitting blood after their packing ashore. This is only a beginning of hard work, which will begin for us Tuesday when we commence our packing over the pass. Some are up to the fourth bench, but as yet none has crossed the summit. Block and tackle has to be used frequently, and 100 to

---

[27]*Excelsior*: This Pacific Steam Whaling Company, three-masted steam schooner met the first down-river boat of the 1897 season at St. Michael at the mouth of the Yukon River. She arrived at San Francisco with more than half a million dollars in gold on June 14, approximately three days ahead of the Seattle arrival of the *Portland*. However, by focusing on the *Portland* and its "ton of gold" theme, Seattle newspapers brought worldwide attention to the Klondike. In 1898, the company put the *Excelsior* on the run from San Francisco via Seattle's Puget Sound to Cook Inlet and the CR. (PR)

[28]*ice creepers*: Tie-on ice cleats for boots. (AP)

150 lbs. is a load on the sled. Dogs are worse than useless, and many have died or been killed rather than feed them.

Tent city this side of the glacier at the head of Port Valdez. The Keystone structure may be the large one, center right. (Snowden Col., Townsend, March 1898)

It is estimated that about 15 - 20 days packing will take us over the glacier. We want to get over before the equinoctial storms. To get caught there, where there is neither wood nor water (snow is melted for water) might be disastrous. Several were badly frozen a short time ago, only escaping with their lives. The weather now seems quite favorable; we had about 8 inches snow day before yesterday. With that exception it has been fair, about 32° during the day, 0° at night. During the day we perspire or get wet from breaking through the snow crust (the snow is 8 - 10 feet deep, with frozen crust upon which we camp); and in the morning we find shoes, socks, pants, mittens, etc., all frozen and have to be thawed out before we can dress.

This is Tuesday morning. Yesterday we had snow all day, about 10 inches, making it impossible to move today. Probably we shall commence tomorrow. Many are packing today, though the trail is in bad condition. Tomorrow we will benefit from this travel as it will be easier for us.

This life seems to agree with me. I am eating enormously and cannot now button my clothes comfortably. Mr. Rayl does not seem to enjoy it. I think he is pretty sick of it all. I feel sorry for him, though; I hear many a silent sigh. It would not require much to make him turn back, I think.

I hope to receive a letter on the *Excelsior* when she comes in; she is long overdue now. Am now writing at 3 p.m. The *Excelsior* came in just as I was writing above, and I immediately went down. She had left all mail at Orca, though, from which point we shall receive it this p.m. The *Excelsior* has about 250 passengers for this point. They are now unloading. This will make about 600 in camp. There are about 8 women. Good order prevails.

The regulations here provide 50 lashes and banishment for thefts less than $100. Over that amount death is the penalty.[29]

Good health prevails generally, though I have had a few calls. At Skagway an epidemic of meningitis prevails, and it is reported that as many as 8 or 10 deaths have occurred in a day. I hope all are well at home. The boys are busily preparing for the packing, which begins tomorrow. I must do likewise and also get the mail to the boat, so will close.

With love to all,
Leroy

Letter 11 Postmarked STEAMER VALENCIA, MAR 19, 98; SEATTLE, WASH, MAR 29, 10 PM, 98; received Beaver Falls April 4

Port Valdes, **March 15, 1898**
Tuesday eve. 8 p.m.

Dear Father, Mother, and Sister:–

We have gotten the most of our supplies to the foot of the glacier. We broke camp yesterday and about 4 p.m. pitched our tents in the city at its base. The day was a bad one for such a move. It was blowing so that the trail was hardly passable, and it was only by lightening our loads or doubling up that we got through. As it was, four of us had to return to the building to sleep. This was tiresome after our pull to the camp above. Today the haul was worse, if anything, than yesterday; and we did not get our personal outfit up, so had to return again. The trip, loaded, requires from 3 - 7 hours. It is about 5 - 6 miles. Tomorrow we shall get up our personal outfit and then commence the still harder work of crossing the glacier. If we get over in six weeks, I think we will do well, though the Col. is counting on less time. It is a hard proposition and requires time. No one has yet crossed with supplies, but one or two have gone light to timber on the other side and returned in about 16 hours. This is quite different, though, from packing over, as many places the incline is such that only 40 - 50 lbs. can be carried.

---

[29] *death is the penalty*: According to Charles Margeson, another CR stampeder, the set of laws that included this regulation was established at a miner's meeting at Valdez. The punishment seems severe, but in that country, provisions meant life or death. (PR)

We were fortunate enough to be able to buy out a party who wished to lighten up and who had goods on the 3d bench. We then sold off our own supplies at the prices paid for the same and in equal quantity and thus got about 1000 lbs. up the glacier 6 miles for nothing. We made a mistake in not bringing a large stock for a store. As it is, we have been selling about $50 worth daily. Flour sells for $10 - $12 per 100 lbs. depending on place on trail. Other things about in proportion. Our camp has been a general headquarters; and Co. K, as we all style ourselves, are well thought of.

We ground leased the first lot of our claim yesterday. Prices on lots $100 inside – $150 corner. 25% down, balance when we can give title. Should gold be discovered on C.R., we will be in it, for everything points to our having the best property around. I hope the commission on surveys may report favorably on this route to the Yukon. If they think well of the glacier route, it will help us; if they follow the river, which enters the bay to the east, we will not be greatly benefited; indeed that would ruin our chances.

We are now striving to have a post office established here with a man by the name of Edwards as P.M. [postmaster]. He will occupy our buildings and look after our mail & goods while we are over the divide, forwarding the first with that of the others as best he can. Until such time as action may be taken regarding the establishment of P.O., letters will be irregular. [30] I have not yet received any since leaving Seattle. They will cost anywhere from 25¢ to $6, though Mr. Edward thinks he can send them over for $1.50 until July as the boats come in. There has been no boat to bring mail since the *Excelsior,* which came in this day week. The *Alice Blanchard,* long since due, has not arrived; nor have we heard of her. Some think she is lost.

Several small sailing boats have come in, but they all left Seattle before we did and bring no news. There are about 1000 people here now. Among the arrivals was a company composed of 34 Connecticut men. They brought with them a freight steam sled, which stands them about $1800 dollars and is a total loss as the sled is wholly impracticable here.[31] How is the Leigh expedition progressing? My short experience here leads me to predict a rough time. Rayl is sick of this proposition, I tell you.

Morning of 16th. The day has broken clear and warm – about 35°; and we will get away in good shape. Am well and taking with me one appetite that paralyzes the cook, as he says.

With love to all,
Leroy

---

[30]*mail and goods ... forwarding ... as best he can*: Dr. Townsend was not the only one to complain about the mail service. When Abercrombie entered the CR valley in 1898, the most frequent complaint was that of individuals who failed to receive their mail, and they expected the "government outfit" to deliver it. Confusion was created when a postmaster was appointed at Port Valdez, but the route was not extended from Orca, the nearest established point. The mail contractors of the Pacific Steam Whaling Company refused to deliver mail to Valdez, turning it over to the local agent for distribution. Finally when the first delivery of the mail by the contractors was made after the military arrived in Valdez in April, a new complication arose. The postmaster at Orca locked the mail pouch and refused to give the key to the postmaster in Valdez. (PR)

[31]*Connecticut men ... freight steam sled ... wholly impracticable here*: This is the Connecticut and Alaska Mining and Trading Association, a party that began with thirty men; it has come to be known as the Margeson Party. Its "ten-horsepower boiler (upright) and ... eight-horse reversible engine" sled was designed "to draw large loads of ... goods over the glaciers...." Including freight, the cost was $2,000 (Margeson 10-12). (AP)

Letter 12 Postmarked STEAMER VALENCIA, MAR 22, 1898; SEATTLE, WASH, MAR 29, 7 PM, 98; [on reverse] received Beaver Falls, April 4, 1898

Glacier City, Sunday, 11 a.m.
March 20, 1898

Dear Father, Mother & Sister:–

I last evening received your letter of February 11. It came on the *Valencia*,[32] which ran into the harbor here Thursday. On account of the ice having gone out of the bay, she could not make landing as we had done and was required to anchor 5 - 6 miles below, at Levanport, there to await the coming of the steamtug *Salino* to lighter goods and passengers ashore. This caused delay in getting the mail.

S.S. *Valencia*. (Puget Sound Maritime Historical Society, # 2601-11)

[32]The *Valencia* was one of three steamers operated in 1898 by the Pacific Steam Whaling Company in service from San Francisco via Puget Sound to Cook Inlet and the CR. This was her first trip to Alaska, an exciting one due to rebellious passengers and a terrible storm. Steerage passengers, dissatisfied with the food, tried to take over the ship, but the captain sprayed them with water from hoses attached to the hot water pipes from the boiler. The fate of the *Valencia* was tragic and ignominious: After a journey from San Francisco to Nome and back in 1905, on January 20, 1906, she embarked for Victoria, British Columbia, and Seattle. On the night of January 22, traveling by "dead reckoning" toward the Strait of Juan De Fuca where she would turn east, the skipper, Captain Johnson, miscalculated and wrecked the vessel against a cliff near Pachena Point on Vancouver Island, a place known as the Graveyard of the Pacific. Over 100 passengers, many never found, drowned or were dashed to death against the sixty-foot cliffs. Over a forty-hour period, ships and tugs that steamed to the area rode about a mile off, fearing to come nearer while the *Valencia* disintegrated in the heavy surf. (PR, AP)

The *Valencia* has about 600 passengers for this place. She came around the Horn from the east, taking on passengers at the various coast ports at which she stopped. Her passage was a rough one, and considerable of her cargo had to be jettisoned. Twenty-four head of horses and cattle with a considerable quantity of hay were thrown overboard.[33] The shipping along the coast in the Alaska trade has been unfortunate this season indeed. This is due to overloading and to the fact that old and unworthy vessels have been fitted out to meet the increased traffic. Then there is great difficulty in getting experienced pilots, particularly on the inland passage. There are not enough who are familiar with the coast to go around. The inland passage is dangerous, too. A hundred feet, 50 feet in some places, from your course may wreck you. We were fortunate in getting off as we did. Had we gone on the rocks at high tide, or had our vessel swung broadside on, we undoubtedly would have gone to pieces. This, too, while we were piloted by probably the most experienced man on the coast, Capt. Lennon.[34] There would have been no loss of life; but it would have been unpleasant, to say the least, to have to have gone into camp hurriedly in such a storm. By the papers received we note the loss of the bark *Canada* near Skagway.[35] The *Alice Blanchard* was abandoned by her passengers at Skagway, we understand, and has returned to Seattle. She is long since overdue here, and much has been the speculation as to the cause.

We are now located at the foot of the glacier camp, Glacier City so called. The proposition of sledding an outfit from Valdes 5 - 6 miles below to this place is a hard one. As I wrote you, too, we moved under anything but favorable conditions. We got along pretty well, though, and are now comfortably fixed. We are still camped on 8 - 10 feet of snow, and today more is falling with every indication of its continuance for some time. While writing the above sentence, I received a call to attend a man who had shot himself in the foot while hunting ptarmigan on the mountainside. I responded, going way up the trail, but found that he belonged to a party with whom there was a physician, so I returned. The wound was not a very serious one, though it may cost him 2 or 3 toes. He rolled 200 - 300 feet down the mountain after the accident. They have taken him to their camp down at Valdes.

There is little sickness here. At Skagway an epidemic of Cerebrospinal meningitis is reported raging.

But I want to speak more of the trail. Like a huge snake it winds northeast from Valdes to the glacier, and from morning until late at night weary packers slowly creep to the mountain of ice in the distance. Wood must be packed from timber about 2½ miles away. The nearest water is ¾ mile off, and thus we depend to a great extent on snow water. A man's load averages 150 - 200 lbs. In crossing the glacier the difficulties increase. There is no wood for 30 - 35 miles. The inclination at some points of the trail is from 10 - 40 or 50°. Tackle must be used frequently, or perhaps I should say is best used. Some, however, pack on their backs, taking about 50 lbs. per

---

[33]*overboard*: The rolling of the ship caused the livestock to suffer broken legs and other mutilations, necessitating their being killed and thrown overboard. (PR)

[34]*Captain Edward Lennan* (correct spelling) was a highly regarded, competent Alaska maritime pilot familiar with the Alaska routes. (PR)

[35]*the bark Canada*: Laden with lumber from Tacoma, the *Canada* stranded near Skagway in a storm February 21, 1898. She was refloated, but in another storm three days later, she was driven ashore and wrecked. (PR)

load. We tackled up the first bench day before yesterday. It is hard work; no one can realize the difficulties who is not present. Yesterday we packed all day to the foot of the second bench, where we cache again. We will place all our provisions there by tomorrow evening, and Tuesday we shall place our tackle and work to the second bench. One who has never seen a glacier can have but a mistaken idea regarding the same. I hope that the film, too, which I send may give you a fair idea of it, though, and the difficulties to be overcome. When we get everything on the summit – about 20 miles from here, we shall cross to the timber beyond – 15 miles – and there pitch camp, packing there from summit to foot.

There is some difference among us as to how much we should take over. The Col. wants to go light and return for balance in 6 weeks or 2 months. This appears risky to the most of us as we have no assurance that the pass will remain open. We are now inclined to take over all supplies to the timber beyond and there make a permanent camp. We then run no risk and can get supplies at any time.

There are no Indians this side the divide. Those beyond are peaceable and will give no trouble. The paper report is a mistake. I wish there were a number of Indians here whom we could get to work for us. With our own labor it will take us 6 weeks or 2 months to get to the timber on the other side with our outfit. Labor costs $10 - $25 per day. As we shall have to go, the Copper River will not be reached short of 150 miles; but the trail after the timber is reached is said to be good. I am told that game is fairly plentiful there, too, moose, caribou, and sheep being found. Good fishing is to be had also. On this side ptarmigan are quite plentiful. Bags of 15 - 18 are common. They are similar to the pheasant of our hills, slightly smaller, and now in winter – are nearly as white as snow itself. Porcupines are found; and a wolverine has made several nightly visits to the camp, each time devouring a considerable quantity of bacon, ham, etc. We have not suffered through his visits, ourselves, though. Cod-fish are plentiful in the bays, but the ice has now gone out, and we have no means of fishing for them. Several fine trout have been caught by the boys while fishing for cod. This happened through their having cut holes at the mouth of a stream entering the bay. They were beauties, though, lake trout apparently and weighed 6 - 8 lbs. When the streams open, I think the fishing would be excellent. I succeeded in getting a Harlin Steel Rod from a party selling out. Quite a number of the boys failed to put a sufficient supply of nerve in their outfit. They have gone back after it. It is one essential if one is to cross the divide.

With regard to the possibilities of the C.R., we are still just as much in the dark here as you are at home. It is an unexplored region. A few men have been on its headwaters, having come in from the Yukon; but we can find no reliable man here who has gone in over the glacier. As I said in my last letter, they have crossed to timber and returned light. The country will receive a pretty thorough scraping this year, though; and by next winter something definite will be known.

Those who come in now will have difficulty in landing. The ice has gone out of the bay; and as there is no dock, everything will have to be lightered in. Whether this will be more laborious than landing on the ice and sledding into camp 3/4 mile away is a question, though.

When we first entered the bay, they wanted to land us at Swanport[36] about 5 - 6 miles below, though it looks hardly a mile. Here the S.S. Companies, expecting to land their passengers at this point, had built a bunk house, though they called it a hotel. They expected to profit through the enforced encampment of the passengers until they could pack across to Valdes. We got the Capt. to land us on the ice, though, and the *Valencia's* passengers have objected so strongly that they will be lightered to the head of the bay. I believe I spoke about distances being so deceptive here. It is astonishing. Even Western men remark at the deception.

I wish you would read or show this to Snyder, Father. I shall try and write him and get letter and films out on the Str. *Moyo,* due here in a few days. How are the films working? If proving fairly satisfactory, start 3 - 4 spools on their way here, addressing c/o F. H. Whitworth, #417 Seneca St., Seattle, Wash. The last spool sent was exposed under more favorable conditions. Have J.A. write me at some length if he can suggest anything. If the films do not work properly, that is, as well as the plates under like conditions, then I think I will try and get a plate instrument in here somehow. I would not take anything for the series of pictures if they prove good. If 50% prove good, I will be satisfied, using films as I have been.

You remember of my having written you regarding the murder and hanging here at Valdes. I have two of the cartridges which were in the murderer's revolver when he committed the crime. One has been fired, and probably the firing of it resulting in the death of one of those killed. Doc. Tanner, the murderer, was promptly hanged.

The order here is good, the men generally being an excellent set.

This will go out on the *Valencia.* I will write you and J.A. in a few days.

Letter will go on the *Moyo.* Regards to J.A. and the boys, etc.

With love to all.

Leroy

[Enclosed with this letter was an additional sheet without heading:]

Sickness – very little.
Troubles – enough to make a minister use strong language.
Dissensions, a few, but of small moment.
Enclose any clippings of interest.
Shall expect more letters on the *Moyo.*
Rayl has clipping stating that T & Co. [Stationers and Job Printers, Beaver Falls, PA] owned by C. W. Townsend, Leroy's Uncle] have let contract for new bldg. This may be after the business was destroyed by fire (PJT).]
Write particulars.
Are all well?
Am getting fat.
Baths are hard to take.
You should see my whiskers.
Monday morning – Snowing, cloudy. Possibly equinoctial storm.
21 March, Leroy
Squibs [short reports, which follow (AP)]
Temperature – Night 0° - 15° or 20°
Day 10° or 12° - 35°
Labor & packing – By day – $10, varies according to what people will pay.
Packing 1¢ to 15¢ lb., varies as above and according to distance, difficulties to be met, etc.
Arrivals – S. *Valencia* Thursday last 500 - 600 people for this point.

[36] *Swanport*: Adam Swan made a location at this site. It is currently the site of the Trans-Alaska Pipeline Valdez Marine Terminal. (PR)

Discharging today.
No. at Valdes and in camp beyond – 1200 - 1500
Prices on goods (What you can get): Flour – $8 - $16 per 100 lbs.
Bacon – 2d grade – 25¢ lb.
Hard tack – 10¢ "
Dried fruits – 20 - 50¢ "
Clothing, Shoes, etc. 2 - 6 times cost in Seattle.
People – Good class, generally, remarkably so, indeed.

Letter 13 Postmarked [ORCA[37]], ALASKA, APR 22 [1898]; [on reverse] received Beaver Falls May 16.

Glacier Camp, Alaska
Wednesday, Mar. 30, 1898
8:50 p.m.

Dear Father, Mother, and Sister:–

As you will see by the above, our party is still at Glacier Camp, or more properly, Camp Valdes No. 2; for it has within a week received this name, officially. How, – I shall explain further on. The storm of which I wrote you last week continued until Tuesday noon, when it cleared up sufficiently to enable us to tackle up the first part of the 3d bench. Wednesday, it was again stormy, and we were prevented from working. Saturday last we had to return to camp on account of the high wind on the glacier. It was blowing a gale, in fact. To-day we are again snow bound; and indeed during the greater part of the night it rained. Thus we are greatly inconvenienced and delayed through the weather.

Lady in the drifts. Snowbound in Glacier Camp, also called Camp Valdes No. 2. (Snowden Col., Townsend, March or April 1898)

However, by hard work Thursday and Friday, we got our supplies to the top of what is known as the third bench. From here the trail, with many turns made necessary to avoid crevasses, leads in a northerly direction some 12 miles to what is known as the foot of the summit.

[37] *Letter 13*: Probably postmarked Orca as ink used is red, as on other letters from Orca; Seattle used black ink. (PJT)

The trail in general to this point is good, though. Then is a constant rise, estimated variously at from 100 - 300 feet to the mile. It is upon this long stretch that we are now working. We have made two trips to what is known as 10 mile camp, sledding from 80 - 150 pounds per man. The round trip constitutes the day's work. We rise at 6:30 or 7 a.m., breakfast, then wind our way up the benches to the top, which in itself is hard work. There we pack our sleds and fall in with the train, which slowly moves along, those in the distance being only black spots and seeming not to move at all. As the various camps or caches along the trail are reached, the boys drop out; some stop at 2 miles, some at 5 miles, others 9 - 10 etc. We reach 10 mile camp about 1 - 2 o'clock and cache there, then return empty, reaching camp about 3 - 5 p.m., tired and hungry. Of course we pack water and lunch with us, but this is not very satisfactory at best.

From the cache we expect to use a horse, which the Col. bought, to sled to the summit about 7 miles distant. Beyond 10 mile camp the trail is steeper at points, but nothing like the benches and ascents to the top of the 3d bench. Quite a number of horses are now being used. They pull 3 - 9 or 10 hundred lbs. depending on the trail. The proposition with horses, though, has difficulties in the way too. If they step to the side of the trail, they break through, as indeed they do from time to time on the trail itself. Last Sunday one got in a crevasse and was with difficulty gotten out. The miners are rather opposed to their use too; then the cost must be considered. Ours cost \$150, and a town lot = price \$150 – in our town site. Hay cost 5¢ per lb., oats in proportion; and we could only get enough for about 30 days short rations. If we can get no more feed, we shall try and send him back to this side and sell him. If this is impracticable, we shall kill and eat him on the other side. There will no doubt be more horses brought in, though, with a good supply of feed, so that this difficulty will be removed.

This is going to be a boom camp, I tell you. Reports say that the Pacific Whaling Co. now have 5000 booked for this place – Valdes – and it is estimated that anywhere from 6 - 40 thousand will come in. I place my estimate in the smaller figures. Boats are coming in every few days. They are mostly of small tonnage, though, and 20 - 60 passengers. Last evening it was reported that the government survey party[38] had arrived in the bay. It has not been verified this a.m. and is probably untrue. All kinds of reports are rife. You must remember this when reading newspaper accounts, too. The transportation companies are booming Copper River, and reports of finds must be viewed with suspicion. We can get no satisfactory information here, and I believe the reports through the States are being highly colored. The C.R. region is as yet practically a "terra incognita." Since I last wrote you, though, several parties have crossed the summit and made camp in the timber beyond. With good weather the next 6 - 8 weeks will place quite a number of strong and determined men in the C.R. Valley, and the character of the country can then be more definitely spoken of. I understand the newspapers report rich discoveries. These must be necessarily untrue, unless something has been accomplished on the headwaters of the Tanana and Copper rivers by miners coming over from the Yukon.

---

[38]*government survey party*: This is Abercrombie's 1898 military reconnaissance expedition, which arrived at Port Valdez April 18 at 6 p.m. (Abercrombie 1900, "ALASKA.—1898," 564). (PR)

X X X X X[39]

The past few days have been eventful ones in the history of the camp. Tuesday the 22d the first death occurred since the tragedy of January last [the double murder by Montana Cowboy]. T. N. Opdahl of Munssenola, Minn., [40] died upon the trail over the glacier of dysentery. Rev. Richards held services over the body; and as these last sad rites were being performed Wednesday morning for the dead, I, assisted by Dr. Barrett, was beginning what proved to be a closing battle with death in an effort to save the life of Leroy Murfin, a young man 22 years of age from Sleepy Eye, Minn. Murfin was shot by the accidental discharge of a gun in the hands of D. Wherley, a young man 18 years old, and of his own party. [41] Murfin with several companions was seated in one of the tents, Murfin himself resting his elbows on his knees. One of the boys picked up a rifle and looking at it, called attention to its rusting. Mr. Wherley, Sr., turned to his son and inquired of him as to whether he, the son, had looked at their gun. Young Wherley replied, "No," but said he would, and getting up went into the tent opposite to do so. A moment later came a muffled report; and Mr. Murfin [the son], rising with a groan, inquired, "Who did that?" and walked outside the tent. A little later the boys assisted him back to the tent. Blood was running down his forearm and hand.

Opposite page: The burial of Leroy Murfin, 22, accidentally shot by D. Wherley, 18, from an adjoining tent, Thursday, March 24, 1898. "Rev. R." (Rev. Richards) may be standing before the grave. (Snowden Col., Townsend, April 1898)

Barrett and I were hastily summoned, and examining we found the left elbow terribly shattered and a penetrating wound just at the upper margin of the Ilium [hip bone]. We worked with him nearly constantly until 2 - 3 p.m. – the accident occurred about 9:00 a.m. – and succeeded in making him as comfortable as could be expected. The greater portion of the bullet we cut from under the skin just to the left of the spinal col. in the small of the back. The finding of part of the bullet and the course which it must necessarily have taken, led us to think that perhaps the abdominal cavity had not been entered and gave us some hope. The injury was due to a 30-30 soft nose, steel jacketed bullet, though; and evidently the jacketing had entered the cavity, for he died Thursday a.m. at 9:20. Young Wherley had gone into the other tent and in taking the arm from its case accidentally discharged it. A case of "didn't know it was loaded." It was a sad affair.

The same evening we held a meeting and elected Dr. Barrett coroner.[42] At this same meeting the name of Valdes Camp No. 2 was officially accepted. The coroner then impaneled his jury – I being one of the jurors; and after viewing the remains, we proceeded to the largest tent in the

[39] X X X X X and sometimes ☆ ☆ ☆ ☆ ☆ are regularly used in long letters for elapsed time or change of subject. (PJT)

[40] *Opdahl, T. N.*: According to newspapers his name was T. H. Opdahl from Marshall, Minnesota. He was buried in his sleeping bag. (PR)

[41] *L. H. Murfin and D. Wherley* were part of the Horace C. Conger prospecting party of twelve men from Minnesota. Murfin's life was insured for $5,000 in favor of his father, and he had $66.33 in his pocket. Five of the party went four and a half miles lower down the glacier to get wood for a coffin, and he was buried on March 25. Sixty-two men gathered at the grave, sang "Shall We Gather at the River," and listened to a minister, likely Rev. Richards of the KC party, read the entire seventh chapter of Revelations. (PR)

[42] Conger reported Barrett was also appointed notary and that the first law for the section was formed at the meeting. (PR)

camp – and there, by flickering candlelight, which made our figures silhouette strangely on the canvas, we listened to the evidence. Outside the wind blew fiercely and a dog howled mournfully with the cold. It was strangely impressive. There was no evidence, however, to show anything other than an accident; and the verdict was brought in accordingly. The next morning at 10:30 Rev. R. conducted services; and the remains were laid in the cold, cold, ground, where the snow is a winding sheet and the whistling winds a dirge.

X X X X X X

Saturday evening a Mr. Stead from Conn. fractured his knee cap. He had gone to the spring for water and fell. There is a physician in their party, though; and he will be well cared for.[43]

Some thieving has been going on. There are so many here now, some of whom are short of supplies. A person found guilty of a theft would receive a severe sentence. If caught in the act, I have no doubt a neck-tie party would be the result. If it continues, I think I may predict a court session with Judge Lynch in the chair. A rougher set of men are coming in, I think.

X X X X X X

I wonder if you receive my letters regularly. I have but one from B.F. since leaving Seattle.

X X X X X X

I have just been witnessing snowslides in the mountains opposite. With a roar like thunder, they slide down the mountain sides – an imposing sight. Every little bit, the falls may be seen.

X X X X X X

The boys have been washing their clothes and writing, reading, etc. today. Two or three have gone to Valdes No. 1. Perhaps they may bring mail.

X X X X X X

Send any clippings of interest from the papers. Enclose some stamps occasionally, lest I run out. Some of the boys did not bring enough, and I have let them have quite a number.

X X X X X X

Tell Al I made a mistake not bringing the 10 inch Stevens Rifle with about a thousand 22 long rifle cartridges. I could have used it on ducks, ptarmigans, etc. nicely; its light weight would make it superior to a shot gun. Shot gun ammunition is too heavy to carry over the mountains.

X X X X X X

As far as my experience permits me to go, I can outfit, I think, lighter and to better advantage than anyone I have yet seen. Each day gives me new ideas.

X X X X X X

It is reported that war has been declared between the U.S. & Spain.[44] I wonder if the report is true. Probably the next boat will have papers. We are looking for the *Alliance* to come in at any time, now.

X X X X X X

The *Alice Blanchard*[45] is now said to have sunk on her return trip to Seattle, 40 lives lost – probably untrue.

X X X X X X

Do not be worried by newspaper reports you may read regarding the camp here or beyond.

X X X X X X

Thursday Eve. – 31st – 98 m. [meridian; i.e., noon]

---

[43]*Mr. Stead from Conn. ... a physician in their party*: This is Lindsey Stead, of Sound Beach, Conn., a member of the Charles Margeson Party; the group's physician is a Dr. Kortright. The injured man, "at the expiration of eight months ... was still at the store, and very lame, though able to be about" (Margeson, 40). (AP)

[44]*war has been declared*: The U. S. declared war on Spain on April 25, 1898. (AP)

[45]*Alice Blanchard*: This schooner did not sink; it was lost in 1929 when driven ashore at West Bay. (*Shipwrecks*, AP)

It snowed nearly all last night but this a.m. cleared away sufficiently to enable us to take the trail. It shortly began to snow, though; and we had a bad day. The trail had to be broken anew, and this caused a delay of several hours. The snow was soft and made walking hard, too. We got in about 4 p.m., tired and hungry. I have since sponged myself off and changed my clothes. They were wet with perspiration and tomorrow morning will be frozen stiff.

The *Excelsior* is in the bay, and we shall probably receive letters tomorrow. She came in last evening and has 230 passengers for this place.[46]

I shall send this out by the *Excelsior* in the morning. Am going to turn in.

With Love to All

Leroy

Apr. 1st, 98, Friday a.m. 6:30. No letters. Will get none until the 12th. Weather good – going to take the trail.

Leroy

Letter 14 Postmarked SEATTLE, WASH, APR 25, 6 PM, 98; [on reverse] received Beaver Falls May 2

Top of the 3d Bench, Valdes Glacier.
Alaska, **April 14, 1898**
Thursday 1 p.m.

I received your letter of March 14 last Saturday. It came on the *Bertha*[47] and is the 2d I have received since leaving Seattle. If any others have been forwarded, they have miscarried. How many have you received from me?

S.S. *Bertha*, in Alaska waters. (Puget Sound Maritime Historical Society, # 328-3, circa 1898-1914)

[46]*passengers, numbers*: Rumors abounded, and there are many versions of the same events. Horace Conger reported 130. (PR)

[47]*Bertha*: This wood steamer plied Alaska waters until July 18, 1915. Grounded in Uyak Bay at Harvester Island, off Kodiak Island, water seeped into the forward hold and ignited a cargo of lime. The Alaska Navigation Company vessel was gutted by the ensuing fire, making it a total loss (Gibbs 72). (AP)

I expected to write Saturday evening last but was prevented through having to serve on a jury – acting as judge – in a case including the division of an outfit in a company of 33, 3 of the members of which wished to withdraw. We took evidence Saturday evening until 11 o'clock and had to sit two or three hours Sunday evening before we reached a decision. It proved satisfactory to both parties. There are many difficulties arising in parties like this.

I wrote to J.A. Sunday. I have received no word from him at all. I wish Lizzie[48] had spoken of the pictures I have been taking. Has Al been getting good results? Perhaps the rolls have not reached him. I am sorry indeed to learn of the serious illness of McMahon and hope it may not prove fatal. Express to Mrs. McMahon my sincerest regret.

I am well, indeed all the party are in good health, though we have been having plenty of hard work and hardship for several days past. I believe I wrote you in my last letter that we expected to get into timber across the divide by the latter part of last week. Our plan miscarried, however, and we did not reach camp until Monday a.m. We got up about 4:30, and from that time until Tuesday a.m. no one lost a minute. We worked, too, on short rations, and a more tired and hungry crowd than we would have been hard to find. Setting up our tents hurriedly, we turned in. Seven o'clock Tuesday a.m. brought a gale of winds, and we had to turn out quickly in order to prevent our tents from blowing down. We got them re-pegged, though, and turned in again, sleeping until the afternoon.

Yesterday morning we started with 1/2 our camp equipage for 12 mile camp. It was snowing lightly when we started. By the time we reached 10 mile cache, however, it was a blizzard; and three of us who were a little late in starting had to cache our loads. The wind was blowing a gale; it was colder, and the trail filling up. I felt some little apprehension as I took the back trail. At times one could not see 100 feet ahead. To lose the trail in such a storm would be unpleasant to say the least. I was wet, too, with snow and perspiration to the skin. Fortunately I had with me a canvas sack which I had used to carry lunch in, and this I pulled over my head. Grabbing my water can and lunch, I started. At the first camp along the trail I stopped, warmed myself, and cut eye holes in my head protection. Here I fell in with two or three who like myself were making back tracks, so I had company. When I got to camp, every stitch I had on me was wet. The other boys followed a little later in about the same condition. To add to our troubles, we were and are practically without wood. Therefore we turned in early and slept late this a.m. Our wood supply is only sufficient to get supper this p.m. We have, however, a considerable supply of food cooked for lunch; and at midnight tonight we expect to break camp and start for the timber beyond. This means hard work until 7 - 8 o'clock tomorrow evening; indeed we will be fortunate if we make camp by that time, for the weather does not now look favorable.

It is dull and cloudy and has been snowing lightly most of the day. The trail beyond 5 mile camp is reported bad, too. As the distance from here is about 25 miles, you can see that the proposition is a big one. The descent from the summit [of Klutina Glacier] can be made in about 2 - 2 1/2 hours, though, as the descent is regular and the trail good. From summit to timber is 12

[48]*Lizzie*: Leroy's sister, Elizabeth Adams Townsend. (PJT)

miles.[49] We shall have about 2000 lbs. to sled, and the hardest work will be in getting this to the summit – 16 miles from here. Tackle must be used about 3 times. From summit to timber 600 - 800 lbs. per sled can be taken easily. Sunday at the farthest, anyhow, will find us in the timber beyond; and if we are favored, we shall sleep there tomorrow night. Our progress then will be much more rapid, I think. We are getting over none too soon either, for the trail below and over the benches is in a terrible condition. I believe many will be prevented from crossing. I do not know what those who are yet to come will do. Many are turning back as it is. Cold weather, though, may favor them. Lately the temperature during the day has been about 40° - 45°; at night it falls to about 32°. It is colder on the other side, of course. Moose and caribou tracks are reported to have been seen there. This is interesting news and will please Al. I think it can be relied upon, too.

None of you need worry about my exchequer. I ran rather short before leaving Seattle – having only about $15 when I left. I have since more than replaced the amount drawn on the emergency fund, having something like $65 or $70. You need give yourself no uneasiness on this score.

As to coming out in the fall, I think in all probability I shall return with the boys; for neither my private outfit nor the provisions will last through the winter. This depends, though, on what the future brings forth. I wish you would send clippings regarding the leaving of the Leigh party. I venture the prediction that they will have troubles of their own. How is Lizzie getting along with her music; is she still attending the Conservatory? Tell me, too, whether you have succeeded in collecting the bills left and what Rouger has accomplished. I will write you from the other side just as soon as possible and try and send a letter back by someone returning to No. 2 camp. I hope the *Valencia* may come in with mail before we get very far away. We are now about to begin packing.

With love to all,
Leroy

Letter 15 Postmarked ORCA, MAY 12, ALASKA [1898]; Seattle May 20, 1898; [on reverse] received Beaver Falls May 25

First Camp – Timber Beyond Glacier
Alaska, **April 22, 98**
Friday 3 p.m.

Dear Father, Mother, and Sister:–

As you will see by the above, we have crossed the divide. It was a hard proposition and we were particularly unfortunate in our moves. I believe I wrote you the 14th, Thursday, that it was our intention to break camp at midnight and go through to timber. I mentioned the weather as being unfavorable at that time, though. It continued cloudy during the evening and early part of the

[49] *Twelve Mile Camp*: so called because prospectors believed it was twelve miles from the summit of the glacier; the U. S. Geological Survey geologist assigned by Abercrombie's reconnaissance team in 1898, Frank C. Schrader, measured the distance at ten miles. (PR)

night, but the boys thought it best to make the effort to cross. I did not favor the move but offered no decided opposition. Accordingly, we arose at 12 a.m. and commenced packing. I had to get into underclothes that were wet with perspiration and water from being caught in the storm before. It is strange what one can go through here and suffer no ill effects. They were fairly frozen when I put them on.

The men of Co. K packing on the Valdez trail. (Snowden Col., Townsend, spring 1898)

We got started about 3 a.m., not stopping for breakfast. All had heavy loads. The first 2 or 3 miles we got along very well, the trail being in pretty good condition. But the wind kept rising; and as we came within sight of 5 mile camp, we found the trail drifted considerably, and pulling became very hard. We reached 5 mile, though, about 6 a.m. Here we heated coffee and beans and had lunch. It now began to snow some, and the wind increased. It was reported, however, that the trail beyond the first mile out of 5 mile was good; and we continued. But this was incorrect; and to add to our difficulties, the storm increased; and we had a regular blizzard to retard us. The trail was soon filled, we were wet to the skin, and our sleds pulled like lead. Fortunately the storm was at our backs, and we slowly made our way along, having to look carefully for the almost hidden trail. We reached 12 mile camp about 2 p.m., wet, tired, and disgusted. The storm, which up to this time had increased, let up some; and we hastily went into camp. A little later it cleared up, the sun shone brightly, and dry clothes and a good warm meal – which we cooked with no little difficulty, for we had little wood, being obliged to break up some of our boxes to use – refreshed us. Saturday we were all too much played out to make any move, and indeed the weather would not have permitted had we so desired. It was cold, windy, and snowing.

Sunday broke clear; and as it looked favorable, we concluded to push on. Accordingly we broke camp at 8 a.m. after a good warm breakfast and pulled on to the foot of the summit 5 miles away. Good weather and a good trail favored us, and we reached there about 12 p.m. We immediately doubled up and commenced to sled to the top. This is the hardest proposition of the pass as we soon found. The ascent is at an angle of about 45°, and it is about 3/4 mile long. From the trail a short distance back, the packers remind you of nothing so much as flies on a smooth white wall. It requires 1 1/2 - 3 hours to make the round trip with a load. A man takes 50 pounds, sledding about 100; but the latter takes longer for the trip. We got 10 sled loads up by 6:30, and about this time ran into a storm on the summit; but we repacked our sleds, pulling the loads of 8 sleds on 4.

Bailey and Rayl, when they reached the top with their first load, pushed right on with a light load. They left the summit about 5 p.m. This left only 4 of us. I should have explained that only six of us are on this side. The others are camped at the foot of the summit and are tackling goods to the top. We are sledding from the summit to camp on this side, 12 miles, Richards, Bruce, Clark and myself, with double loads. It was 7 o'clock; the storm had increased. The trail was filled with snow, and it was blowing hard. We thought, though, that as it was all down grade, we could make it nicely, and off we started. We were again in error. On account of the condition of the trail, it was a dead pull all the way. On two of the steepest places only did we realize that we were going down grade at all. The most of the trail we could only feel for; it was impossible to see it, and we took turns in breaking the way. We were without water, too. For 5 hours we struggled on, wet with perspiration and snow; yet it was freezing cold. 12 o'clock just brought us to the foot of the glacier – 8 miles only. But here we found water, and I think none of us ever enjoyed a drink more. I never did, I can tell you.

We were too tired to go any further and decided to camp. Bedding we had only for three. Rayl had taken Clark's bed with him. But we threw our tent over a couple of loaded sleds, cached there, fastened it down with one of our own to windward, and crawled under. We spread out our bedding as best we could and turned in, having to wedge ourselves like sardines. For lunch we had hard tack and molasses. Even under these conditions – wet, with a freezing wind blowing, ice and snow for a mattress, little to eat, and insufficient bedding, I got some rest. Luckily I was not an end man [in the makeshift tent].

We got up about 7 a.m. and made the camp below about 12 m., where we got a good meal at the tent of a Mr. Starweather. Bailey got us the meal, as he and Rayl got through. They had lighter loads and an earlier start. We set up our tents during the p.m. and turned in early. Tuesday we got our camp in order and went for wood. Thursday we packed wood to the boys at the foot of the summit, the weather being good. We left at 6 a.m., and I got in at 10 p.m. I started back with 550 pounds but had to cache and come in light. Richards played out and could not return. Rayl got through with about 450, but he got an earlier start. I would have gotten in but had to quit on account of Clark, whose eyes failed him. Snow blindness is very common.[50] None of us has been seriously affected, however. Yesterday I went for the load I had cached the day before and brought it in. The trail was good. Today all have been too tired to work, and we have remained at camp – washing, writing, chopping wood, etc.

I have received no more letters yet, but we are expecting the mail carrier at any time now.[51] The *Valencia* brought in a large amount a few days ago. Letters cost $1, or rather one or more at the same time costs $1 for delivery.

Sunday p.m. Apr. 24, 98. The mail carrier came yesterday p.m. but brought in no letters. Our package has miscarried. He had to deliver about 500 letters. Today he has gone on to the lakes 15 miles beyond. Between the farther shore and this camp there are about 200 persons now. The lake is about 5 x 20 miles. It is still frozen over, and all are making the greatest haste possible with the hope of getting across before the ice breaks up. The trail between here and the lake is soft in places and will no doubt cause us trouble. We expect to have all our outfit at this place by the middle of the week and will immediately move ahead. Moose and bear tracks have been seen. The second river has been reached = [per] report. This necessitates crossing another divide of some 10 or more miles, but 200 pounds can be taken on a sled here to the man. The government expedition is now at Valdes No. 1 under Col. Merriam, I believe.[52] The trail between this point and the 3d bench is about destroyed. No sleds can be used. 200 argonauts came in on the *Valencia* the 18th or 19th inst. These will have a hard time. Many are turning back. Will write again soon. All are well. We miss the papers greatly and are ignorant of the outside world as though it did not exist.

With love to all.
Leroy

---

[50]*snowblindness*: Exposure to the sun's ultraviolet rays, reflected from the ice and snow, causes a painful eye inflammation that usually results in temporary blindness. The prospectors used a variety of devices to avoid snow blindness, and many adapted the Native's close-fitting goggles carved of soft wood. The inside surface was blackened to reduce glare from the small amount of light admitted through tiny eye slits. Snow blindness generally took three to four days to effect a cure. (PR)

[51]*mail*: The postmaster at Orca made arrangements with Peter Jackson (identified in other accounts as a "Squaw man," a term used to designate a white man married to an Indian) to carry the mail from Valdez into the Interior. Jackson, with a companion, attempted to find a more direct route from Valdez to the various camps he served. He finally succeeded, but not before his companion perished from freezing, in reaching the Lowe River valley through a divide. When Abercrombie's expedition landed in the spring of 1899, Jackson's discovery had become generally known, and several parties of prospectors who landed early in the season were already at the summit of the Lowe River divide, bound for Quartz Creek with their season's supplies. This is the route of today's highway from the interior to Valdez. (PR)

[52]*Col. Merriam*: Townsend recorded a rumor that was incorrect. Captain W. R. Abercrombie was in charge of the U. S. government expedition to locate practical routes of travel from tide water to interior regions. Details of this expedition are given in the historical introduction. (PR)

Letter 16 Postmarked ORCA, MAY [day illegible, 1898]; Seattle May 20; Chicago May 24; [on reverse] received Beaver Falls, May 25, 1898

First Camp – Timber beyond Valdes Glacier
Friday, Apr. 29, 98, 2:40 p.m.
Alaska

Dear Father, Mother, and Sister:–

The mail carriers passed through several days ago on their way to camps beyond. They had no mail for us. Either ours has miscarried or has been purposely held by the agent of the Pacific Steam Whaling Co. at Orca, who is Post Master. It has been suggested that they feel displeased with the Keystone Co. inasmuch as the town-site located by us has appeared so far to be preferable to their location. I am of the opinion, however, that our mail has simply been overlooked. The carriers are here now on their return trip and are being entertained by us. They have been with us since last evening and will remain until tomorrow a.m., possibly longer, as storms have rendered it impossible to cross the summit. The wind is now blowing a gale. Half of our party is still on the other side. We have had no word from them since Monday last, at which time they had wood to last until Wednesday only. They are probably existing on uncooked rations, unless through good fortune they have been able to buy wood. For three weeks it has stormed almost daily. The Valdes pass is worse than the Chilkoot. Many will be caught by the thaw, now beginning, and will have to return. The trails are getting worse and worse. We have got to make 50 miles yet before I shall feel perfectly safe. This will place us on the second river, a stream navigable for small boats and emptying into the C.R. This we expect to do without serious difficulty, though. This places us beyond the coast range of mountains and in the C.R. Valley, though the distance to the river is as yet unknown. Here the character of the country changes. It ceases to be mountainous. To the west, north, and east – as far as the eye can see – extends timbered lowland. It is said, though, that Mt. Blackburn and Mt. Wrangle have been seen in the distance.

The C.R. government expedition is now at Valdes. There is no mistake this time. They have taken the Pacific W. Co's 160 acres and the greater part of our town site, too.[53] This is about all we know of the matter, though. There are 37 soldiers. It is presumed, however, that anticipating the Mail Route through to Klondike from this point, the Government ordered that a section be held as a terminal, and that the Commandant in Charge simply followed orders and took the most desirable lands. Our application was forwarded about Mar. 10. If it has been received and recorded – we have not heard – this may be greatly to our benefit; for the claim will undoubtedly

[53]*our town site*: The military reservation selected by Abercrombie and used that year and in 1899 was found to be utterly valueless for any purpose, owing to the overflow from streams formed by the melting ice of the Valdez Glacier during the summer. Abercrombie moved to a different location called "Ludington's Landing," which became the site of Fort Liscum. (PR) There were apparently other troubles regarding the site as well. In a letter to Burger Hermann, U.S. Commissioner General Land Office, in Washington, D.C., draft dated May 13, 1899, Mining District Recorder J. F. Kertchem complained of a dispute over the district's authority. "Now further," Kertchem writes, "in the last two weeks some parties have jumped the claim the 'Keystone' company (a trader's claim) and have organized a town com[mittee] and have surveyed same in part and they have voted our claim into the town limits...." (From a typescript of the letter supplied by Nancy Lethcoe.) (AP)

hold and proper remuneration be made. The P.W. Co's notice is defective, so their claim may not hold good. We have complied with the law in every way.

Will write again in the near future. I wish I could hear regarding the pictures sent J.A. I hope they are proving good and that some rolls [of film] will soon come to hand, together with letters that I know have been written.

With love to All,
Leroy

Letter 17 Postmarked [only] ALA [of *Alaska* is clear]; [on reverse] received Beaver Falls June 18, 1898

1st Timber Camp[54] – Beyond Valdes Glacier
Alaska, **Sunday, May 8, 1898**
2 p.m.

Dear Father, Mother, and Sister:–

Yesterday p.m. I received a letter from J. A. Snyder together with a package containing four rolls of films for camera. The letter was dated Apr. 4th and contained quite a number of clippings. I was glad to be in receipt of same, indeed. I do not know why I have not received your letters; only two have come to hand. J.A. speaks of Mother having called on Mrs. Snyder and speaks of her as looking well though being somewhat discontented. The others, he says, are in good health too. I am glad all are well. I am standing the trip first rate and have gotten fat even though the work has been terribly hard. All of Co. K. are now in camp as above. The boys who were on the summit got down Thursday last. They had a hard time on the summit. For one week it stormed incessantly. During a great deal of the time it was impossible for them to go out of the tent. About 7 feet of snow fell. It covered tents, caches, tackle, etc. and in drifts measured 10 - 18 feet. Many caches were lost entirely; and I know of nothing that affected me more than to see poor fellows, with desperation and despair written on their faces, "prospecting" for lost stores. We lost 150 - 200 pounds of flour and about 800 feet of rope, but consider ourselves lucky, indeed. We have our tackle located, but as it was laid, it will require us to dig a trench throughout the greater part of its length, 800 ft., and to a depth of 8 - 10 feet. We shall make an attempt tomorrow with windlass and pulley to get it out. The boys had to remain in bed much of the time, and you may realize their discomfort, for they were wet all of the time. They were short of wood & they had no water. One day all our party had was 2½ hard tack each with snow to wash it down. Others were even worse off. At the foot of the summit the difficulties were equally as great. Two snow slides occurred there. In the first about 15 were covered, two deaths resulting. In the second 5 or 7 with no deaths. The first fall occurred Saturday the 1st at about 10 p.m. At this time our boys were at the foot, but they were about 200 feet lower down. After the slides the rescuers worked almost incessantly for 48 hours. Slides are also reported from the 1st and second bench with loss of life, but we believe these reports to be untrue.[55]

[54]*1st Timber Camp*: This was more often known as Twelve Mile Camp. (PR)

[55]*snowslides*: This is the blizzard discussed in the historical introduction, which resulted in a series of avalanches. (PR)

Spring is fast breaking, and more trouble may be expected for those on the glacier. I am thankful we are off. Most of our outfit is cached 7 miles below here; and if the weather is favorable, we shall move camp Tuesday. This will place us within about 8 miles of the lake. Here we shall remain until we are enabled to raft down the river to the lake. Possibly we shall be able to continue right on to C.R. [the Copper River]. The beauties of the country increase as we proceed. Game will be found in plenty, I think. We see ducks and geese almost daily in their flight now. Moose, bear, and mountain lion tracks have been seen, too.

The report of the soldiers having taken our townsite has proven a mistake, and yesterday we received our notice of "Application for Survey" from the U.S. surveyor general. This makes our claim to the land all right. I hope it may prove valuable. There are 10 or 15 buildings on the property now, the nucleus of a much larger place, I trust. If strikes are made beyond, we shall have a good thing without doubt. Many are turning back. Just now I learn that the Col. is dickering with a Dr. Cunningham relative to buying his outfit. There are many whose hearts have failed them.

I am pleased to learn through Al that the pictures are giving good results. They give some idea of the proposition before us. The next roll will contain pictures of the camp on the foot of the summit before and after the snow slide, as well as the summit itself when the loss of so many caches took place. I hope they prove good, too. Did you receive a photo of the party from Seattle? And have you received any paper with account of interview of self and others of Co. in it?

Occasionally address me Port Valdes, Alaska, only.[56] Write frequently and enclose any clippings of interest.

We have papers of Apr. 16 - 17. Everything is war, isn't it? Great interest is shown all along the trail.

With love to all,
Leroy

Letter 18 Postmarked O[RC]A, ALASKA, MAY[day illegible, 1898]; [on reverse] received Beaver Falls June 18, 1898

Camp – 6 miles below 1st Timber Camp
beyond Valdes Glacier, Alaska
Sunday 7:30 p.m. **15th May, 98**

Dear Father, Mother, and Sister:–

Mr. Reed, a friend of mine, and just returning from Copper River, is spending the night with us. He is going back to Valdes to buy footwear and provisions and will carry this through to that place. Reed has gone through light to C.R. and gives us much information of value regarding the trail and country. Spring is upon us; the snow is going off rapidly, and sledding is now about over. Reed reports the snow all gone in the C.R. Valley. Fortunately we have reached a point from which we can boat to the Copper R., and tomorrow we begin the construction of 3 or 4,

[56] *Occasionally address ... only*: It seems LST is requesting that a basic, simple Valdez address be used as opposed to the forwarding address in Seattle because as he says above, "I do not know why I have not received your letters...." The Orca postal authority dispute may also be a reason for identifying Valdez as an address. (AP)

thirty-foot boats. I believe I wrote of the route we anticipated taking – sledding to the lake,[57] crossing on the ice, making a portage to the second lake,[58] crossing it on the ice and then sledding over a divide to the 2nd river down[59] which we expect to boat, a distance of 80 to 90 miles in all.[60] This is the route which Reed took and returned, and the distance noted above is his estimate. Our plans are now changed. From this point we shall boat down the stream we are on to the lake about 8 miles below.[61] The lake is about 25 miles long. From its foot a small stream[62] empties into C.R. This has been a well known fact, but it has generally been understood to be filled with rapids and falls. Exploration has proven this not to be the case and shown it to be a feasible route. Quite a number have already gone down this stream. They, however, went down on the ice; but the stream is now opening up; and soon boating will be possible. This stream, which according to Reed, who followed the bank from the 2nd River[63] to its entering the Copper, rates only 10 miles lower than the 2d.[64] It saves 25 or 30 miles of portage, though. This, of course, will be the route followed; and if our good fortune continues 2 weeks more, should place us on that mysterious river.

I tell you we have had hard work making this point with our outfit. We have worked day and night the past weeks, sledding over trails through which we slumped to our waists, wading through water which frequently found its way over our hip boots. But all this is now over. We are situated better than we have yet been. Our caches are within easy reach, and tomorrow a.m. we begin to whipsaw lumber for our boats. I am sorry for the poor fellows who are still working on the glacier. For the past few days it has been storming there, too. This glacier proposition is a hard one. The Chilkoot pass is no harder, no more dangerous, I think. I have seen much to bring a shade of sadness and sorrow: gaunt and hungry dogs, starved horses, distressed men. Of the men not a few have told me of their troubles. I have lost no opportunity to extend a hand or do a favor, and I think I may say I have done much to gain for Co. K the esteem it commands.

Our property at Valdes is all right. Moose, bear, caribou, and mountain lion signs are plentiful at and beyond the lake. Flowers in bloom in C.R. Valley. Robins, flickers, swallows, and other birds of Beaver Co. make it homelike here. Will write to you and J.A. again soon. Am sending him another spool of films with this.

What is the war news?

We expect the mail carrier anytime now, and I hope for letters from you.

With love,

Leroy

All here in good health and cheerful.

---

[57] *the lake*: Lake Klutina. (PR)

[58] *the second lake*: St. Anne Lake. (PR)

[59] *the 2nd river down*: Tazlina River. (PR)

[60] *a distance of 80 - to 90 miles*: A prospector's trail already existed over this route. (PR)

[61] *the lake about 8 miles below*: Lake Klutina. (PR)

[62] *a small stream*: Klutina River. (PR)

[63] *followed the bank from the 2nd River*: Prospectors believed the second river was the Kosina (called Tiekel on today's maps). This really was the Tazlina River. (PR)

[64] *10 miles lower than the 2d*: The Klutina River. (PR)

Letter 19 Postmarked [OR]CA, ALASKA, JUN 10 [1898]; [on reverse] Seattle, June 22

In Camp, Valdes Trail, Alaska,
6 miles beyond 1st timber
Thursday, May 26, 98, 2 p.m.

Dear Father, Mother, and Sister:–

I wrote you a short letter Sunday, the 15th inst., the day after having pitched camp as above. I hope it will duly come to hand. The mail carrier passed through here Tuesday – m. [meridian] but had no letters to me from you. I hope all of my letters are being received; otherwise I know how anxious you would be. Mr. Rayl had word from home, however, and tells me that Lou had called on his wife and suggested as a cause for my not hearing from you, that the impression Lou and perhaps all of you have that the letters are costing $1 - $2 for delivery may have deterred you. Of course such is not the case; but I now remember having written you that mail would cost $1 for one or more letters delivered at one time; and I am surprised to think that I did not make explanation later of the arrangements we had been able to make with the carrier whereby, in consideration of a town lot at Valdes No. 1, he was to carry mail of the Co. without any further expenses. It escaped my attention, however. Mr. Rayl tells me that his wife knew of these arrangements, though; and undoubtedly she told Lou; and thus you have learned of them before this. I shall look for a number of letters in the next delivery. I should like to have any clippings of interest, too, and an occasional Dispatch for the war news. We have San Francisco papers of May 31, and this p.m. a party told the Col. that he had seen a paper of the 15th or 16th and that our victory in the Philippines was a fact.[65] All along the trail the greatest interest is manifested. There is some of the material here, too, of which soldiers are made.

We now have all lumber sawed for our boats, and one of the latter is just about completed. The boys are just now putting on the finishing touches. If the fine weather which we have been having continues, a few more days will find all completed; and our voyage to the river will begin. Mosquitoes are becoming very annoying. They are great big fellows; but in proportion to the size of the Beaver Co. "cutter," they are not in it. They will probably become more troublesome as the season advances, though. Reports from the front of the line are not very encouraging, particularly along the line of the trail that turns to the left at the first lake. As the trail has broken up, the men have found themselves in swamps or on unnavigable streams. They have had a hard time; and many, becoming disheartened, are turning back. Mr. Rayl and Mr. Henning, of our Co. have decided to give up the struggle. They will return by the first boat possible, and Rayl will probably reach B. Falls as soon as this letter. He is not at all fitted for it and from the first has been disheartened. As far as the reports from the river are concerned, we can place no reliance on them at all. Discouraging reports have been made all along the trail. But all obstacles so far have been overcome, and the probability is that those of the front will likewise fade away.

---

[65]*victory in the Philippines*: The U. S. declared war against Spain on April 25, 1898, and Commodore George Dewey entered Manila Bay in the Philippine Islands on May 1 with an American fleet. The entire Spanish fleet there was destroyed in one day without the loss of a single American vessel. Spanish forces surrendered July 17, 1898. (PR)

Timber being cut for boat building, Valdez Trail, 6 miles beyond First Timber (Sawmill Camp). A bemused Leroy Townsend looks into the camera, center right. (Snowden Col., May 1898)

As I wrote in one of my first letters, the C.R. region is practically unknown. Probably the number of white men who have done any exploring on the stream does not exceed a dozen or twenty, hence the difficulty in obtaining reliable information. The number of Indians in the Valley probably does not exceed 300 - 400. These cannot be relied on either. The next three or four months will settle the question, though, and the region no longer be a mystery.

Friday a.m.. Arrangements satisfactory to the Co. and to Mr. Rayl and Mr. Henning have been made without difficulty. They receive $100 each in cash and retain 1/2 of their proportionate interest, a twelfth, making a 1/24 interest that they will hold in the property at Valdes and in any location made beyond. On the other other hand, their share of the outfit becomes ours. This is a very satisfactory settlement, and there are <u>no regrets</u> expressed at the withdrawal of the gentlemen, both of whom have been a source of irritation. Rayl, particularly, has made himself disliked.

This morning encouraging reports come to us through a friend who is camped midway on the stream flowing from the lake to C.R. According to him we come out on the river above Wood's Canyon and within a short distance of the Chittyna River.[66] His party gets this information from

[66]*Chittyna River*: Today's Chitina River. Wood's Canyon, several miles long, is where the CR enters the Chugach Mountains. The narrow gorge required a portage. (PR)

the Indians. We have not expected to strike C.R. so far up and hope this information is correct. He also reports many are losing their outfits in going from the lake to the river, the result of careless handling their boats. They are in too great a rush. The water is rising now, though, and less difficulty is experienced. Those who have built scows seem to have been the most unfortunate.

This p.m. Mr. [O. A.] Richards will hold services at the funeral of Mr. Grund,[67] who died in this camp Wednesday evening. Death was the result of scurvy. The case was remarkable on account of the early development of the disease, as the deceased had only been away from San Francisco about two months. He was a patient of mine, but unfortunately I did not see him until weakness made recovery impossible. He had been waited upon by another M.D., who evidently did not recognize the disease and apparently treated the case as one of gangrene of the legs and feet induced by too tightly constricting the parts with ropes in tying on sacking to prevent slipping on the snow. This is the first case of scurvy I have seen. There is little sickness along the trail.

I spoke of the snowslide in my last letter, resulting in the death of two men. I have since learned that one of three dogs buried was taken out alive a week later – rather remarkable.[68]

Sunday – May 29, 98. 10:00 a.m. Rayl and Henning left this morning on their return to the States. They departed sooner than anticipated, hoping to catch a vessel at Valdes which we learned was in. Should they do so, Rayl may reach B.F. before this letter. Rayl has gained for himself the contempt of the whole company, not alone on account of his return, but because of his mean disposition. Of course he has been greatly depressed, and this must be taken into consideration. Undoubtedly his depression will cause him to paint the conditions here much darker than they really are or have been, so you must make allowances.

Monday – May 30, 98. Mail was delivered to Co K. by a friend coming from Valdes, but I was not among the favored. In a B.F. *Tribune* received by one of the party, dated May 2d, I note with sorrow an account of the death of J. A. Tatro. Poor J.A. gone. I can hardly realize it. I must write Mrs. Tatro expressions of my sympathy and regret. I presume his affairs were in about the same shape they were when I left. Al stated in his letter to me that nothing had developed up to that time – Apr. 4. The paper is full of war news, too. Dr. Coffin has volunteered I see. If I were home, I should do likewise. We rec'd Seattle papers up to the 17th. Upon their receipt all work was stopped and their contents read aloud. Don't fail to send a Pitts. [Pittsburgh] paper.

May 31st 98. Today three of us have been working at the saw-pit getting out some lumber for seats, oars, etc. We have our second boat about completed and a third well under way. Weather warm and pleasant and everything moving smoothly. Have you any news of the Leigh party? I presume they have reached St. Michael by this time. Finds are reported from the Tanana and, too, from Kotzebue Sound. We ought soon to be getting reports of prospectors on the Copper.

Wednesday, June 1st, 98. We are having beautiful weather and are pushing the boat building. Today I took pictures of the camp and shipyards. These will give you an idea of our

---

[67] *Mr. Grund*: Horace Conger wrote that August Grund was 38 years old. (PR)

[68] *remarkable*: C. Margeson describes how a man's dog team was buried in one of the snowslides and after all the people were rescued, the owner began to dig for his dogs. Each of the first three were dead, but the fourth was found alive many hours later. (PR)

surroundings. We learn that Rayl and Henning were too late to catch the Str. *Moyo* as they had hoped to. They may be delayed several days or even weeks.

Thursday, June 2d, 98. Work this morning was continued on the boats, I being engaged in caulking one of them. About noon, however, a high wind arose, so strong indeed, that we have not attempted to work this p.m. There now seems to be no longer doubt as to where we shall strike C.R. Everything points to our coming out above Wood's Canyon and the Chittyna River,[69] as the Indians say we do. I enclose a rough map [see next page] which shows our probable course. Instead of taking a course eastward from Valdes No. 1 along dotted lines to □,[70] we have constantly traveled north, even north by west; and thus we reach the Klutena River,[71] which empties into the Copper just below the great bend. We all hope this shall prove the case. Indians have not yet come up this far, but they camp at the lake below. They say the salmon run will soon begin. I would enjoy fresh fish or meat, I can tell you. A small lake a short distance below us here contains plenty of trout, but we have not succeeded in catching any as yet. Bear sign are plenty around the lake, too. One was killed 3 miles above here a few days ago. Within a stone's throw of our tent, beaver sign is plenty; and a few hundred yards below is evidence of their work in a dam.

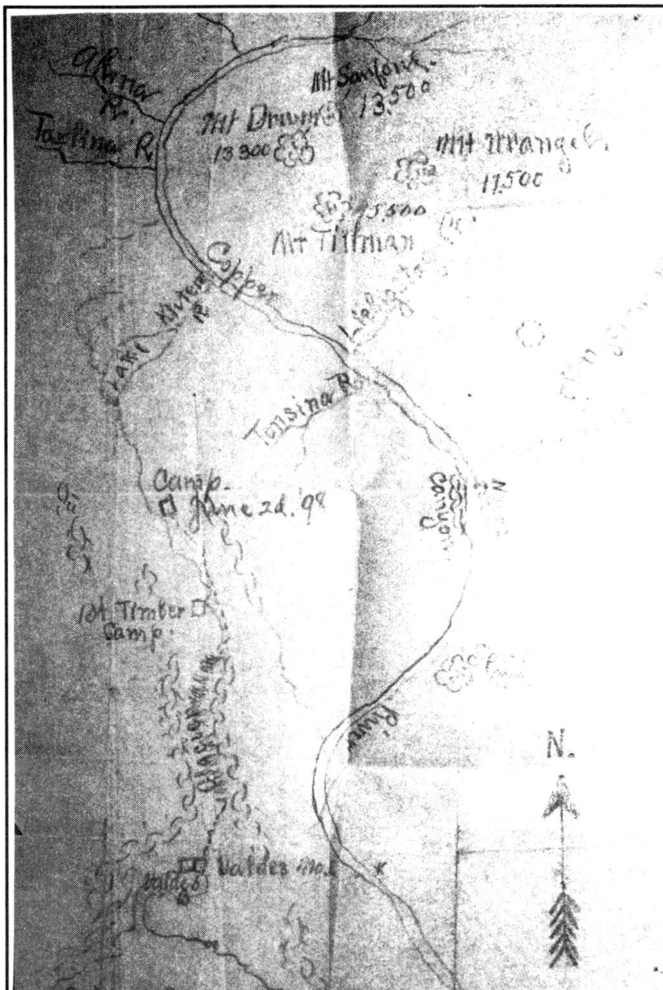

Map from Valdez into "terra incognita," as termed by Townsend (p. 55), enclosed with Ltr 19. (The Conger map, p. 22, provides a clearer perspective.) (PJT Col.)

Friday, June 3rd, 98. Today we practically completed 3rd boat and commenced the 4th, which is well underway. We finished caulking the first and launched it. It rides the water beautifully. The dimensions are as follows: length overall 30', keel 25', beam 6', bottom 3'. It will hold about 6000 lbs on a 12" draught. An opportunity offers to send this to Valdes tonight, and I will take advantage of it. Tell Al of its content.

With love,

Leroy

Opposite page: Boat building operations and the camp, Valdez Trail. (Snowden Col., Townsend, June 1898)

[69]*Chittyna River*: Today's Chitina River, a river nearly as large as the CR. (PR)

[70]□: The June 2 camp site, probably Sawmill Camp. (AP)

[71]*Klutena River*: Today's Klutina River. (PR)

Completed shallow-draft boats beside the Klutina River. (Snowden Col., Townsend, June 1898)

Letter 20 Postmarked O[RC]A, ALASKA, JUN [day illegible, 1898]; [on reverse] Seattle, June 22; received Beaver Falls June 28

Camp – 7 miles beyond 1st Timber,
Valdes Trail, Alaska
Tuesday, June 7, [98], 7 p.m.

Dear Father, Mother, and Sister:–

I wrote and forwarded you a letter several days ago, which I hope you will receive promptly. It was sent to Valdes by two men who were returning to the States. Peter Jackson, the mail carrier, now on his return to Valdes, will be entertained by us tonight. His reports from Copper River are quite favorable. Gold is being found on the bars, though not yet in any paying quantities. As the mountains to the north are neared, better paying dirt should be found, however. According to him every bar shows colors. On the Tazlina River several claims have been located. This is the river next above the Klutena, down which we boat to C.R. We are made happy in the absolute assurance, too, that we strike Copper River about 60 miles above the Chittyna, and so, too, above Wood's Canyon. This should make our property at Valdes of more value, for it decreases the difficulties in the way of an all American route to the Klondike via Copper River.

The dangers of boating from the lake down are evidently greatly overestimated. Many are getting through safely without, as Jackson puts it, "even wetting so much as a bean." Care and a little time is all that is necessary. Co. K will take both. We expect to start for the lake tomorrow and hope to proceed to C.R. without setting up camp. All our boats are completed and afloat. The weather is favorable, though for a day or two high winds have been blowing and rather chilly.

We are hardly having any night at all. At 9:30 the setting sun lights up the snowcapped peaks in golden colors; and at 1:30 or 2 a.m. a glow in the east heralds the approach of morn while a moon, larger and brighter by far than you have ever seen, lends beauty and enchantment to an Alaskan night.

Jackson will be through again from Valdes about the 25th or 26th. I shall be disappointed indeed if he does not have letters from you. Write me regarding poor J.A.'s [J. A. Tatro's] death and burial and how Mother, Sister, Grandmother, Lou, and indeed all, are. Remember me to Snyder, and friends in general. I am sending Al under separate cover a spool of films. Tell him to take special care in developing it, a picture of Rayl and Henning[72] as they began their homeward journey being particularly of interest to the boys. By the way, let me know what Rayl's "tale of woe" is. It will be interesting too. Tell Al what I have said of him and speak of the picture, No 6 on the spool, I think. I am surprised that the pictures so far have not created more interest.

Hoping all are well, I am

With love,
Leroy

Letter 21 Postmarked SEATTLE, WASH, JUL 18, 2-PM, 1898; [on reverse] received Beaver Falls July 23, 1898

Lake Camp, Valdes Trail, Alaska
June 13, 98, 9:15 p.m.

Dear Father, Mother, and Sister:–

I have almost lost all appreciation of time this past week. Day has been turned into night; or rather, there is no night; and we have been working the whole day long. Thursday last we broke camp at 7 mile and started in our boats down the river for the lake. We left 7 mile about 12 m. and reached the lake about 7 p.m. safely. The delay was due to our having to unload and handle our cache, which was located about a mile below the camp. We had no serious mishap. Several times we ran aground and had to get out and push off, and boat No. 1 lost a couple of shovels through having them dragged off by overhanging brush. The stream is swift, and its many abrupt turns make it somewhat difficult to run. At least many find it so, for every day some poor soul is wrecked. All along the shore goods are seen that have been fished out by the kindly disposed and left to be claimed by the owners. I happened to see a boat upset and took a picture or two.[73] They will come in the next spools.

We are now camped in a most beautiful spot on the lake. I wish we might remain for some time, but we shall probably start across the lake day after tomorrow. We have just received word that the surveyor is at Valdes to survey our property. This necessitates the Colonel's going back to Valdes, and he expects to start tomorrow a.m. The round trip will require about a week. We shall proceed across the lake without him. He will take this note with him, and I hope on his return trip to be favored with many letters from you.

[72] *picture of Rayl and Henning*: This photograph is not known to exist (AP)

[73] *boat upset ... picture or two*: This photograph is not known to exist (AP)

The mosquitoes are terrible. As I write this, a smudge is burning in the center of the tent just before me; and my head is done up in mosquito bar; yet they are most annoying. Mr. Richards is overhauling his guard that he may get some sleep tonight. I must do the same. Mr. Warren couldn't stand the smoke, so he has moved outside where he has fitted up a guard.

One can sleep out now without any trouble. Indeed, the weather is delightful, though it seems strange with the snowcapped mountains all around. The snow is fast disappearing, though. All are well and in good spirits. I will write again soon.

with love to all,
Leroy

[On the back of the map enclosed with this letter, dated June 3d, is the following:]

The two men through whose kindness this will be taken to Valdes are just from C.R. Sickness causes their return. One apparently is developing scurvy. They report that we are 60 miles above the Chittyna. The Indians tell them that white men are getting gold on the Chittyna and, too, about 40 miles above on the C.R.; that is, above where we strike the Copper.

I wonder what the tide of war is.

Leroy

[Also on the back of the map is the following:] Leroy S. Townsend, M.D.
Valdes Trail – Alaska June 2d, 98

Men of the Keystone Company in camp, wearing their "mosquito guards." (Snowden Col., Townsend, June 1898)

Letter 22 [No envelope]

In Camp, Amy's Landing.
Valdes Trail, Alaska
Tuesday, June 28, 98

Dear Father, Mother, and Sister:–

I wrote you two weeks ago from the head of the lake; and the Col., going back to Valdes, took the favor with him. I hope you will receive it in due time. Since then we have moved down the lake and the first four miles of the Klutena River to the above point, Amy's Landing, just at the head of the rapids.[74] We made the trip safely, breaking camp Saturday the 18th at 11 a.m. and reaching here at 10 p.m. The distance is about 24 miles. The ride on the lake was enjoyable, though it was intensely warm, and until 5 o'clock not a breath of air stirred. This made rowing seem like work, for all of the boats were heavily loaded. About 5 o'clock, though, a stiff breeze sprang up and helped along considerably. I did not benefit as much as the crews of the other two boats. They each had a sail; I had none. I took advantage of the breeze as much as possible, though, by tacking. But the waves were running so high that a good deal of the time I had to run with the wind to keep from shipping water, considerable of which we did take aboard. The other boys had the advantage of larger boats, 30 feet, and 3 men to a crew; while I, with Julius Broke, a protégé of Co. K., was in a 25 foot boat. In about three hours we reached the foot of the lake. A landing was made by the other two boats that were in advance of me. The mosquitoes were simply unbearable; and Warren, learning that the first four miles of the river was good, thought it well for us to run down. The river is good for this distance as compared with the rapids below, but this recommendation need not place it in very high favor. Warren and one of the others were soon badly scared and "wished they had been good"; and all of us, indeed, were more or less apprehensive; for, not anticipating the run, we had not tied our cargoes in. Sharp was thrown out but with no damage further than the wetting. His boat, too, went over a big rock in the channel, a dangerous one; but the good fortune of Co. K. favored. I got through without any incident worthy of note. In handling our boat I have always been at the oars and have run stern foremost, which I deem the best way. Thus you can always see the water ahead and, by maneuvering, pick out the best channel. By backing water, you can get steerage away, if desirable; and in case of necessity, by pulling, you can naturally lessen your speed. The Klutena from here on is a miniature Niagara. Seven out of ten who have undertaken to run it have come to grief, and along its shores everywhere may be seen goods drying.[75] The current is probably 15 miles an hour. Everywhere are hidden rocks, and the turns are so abrupt that a hundred yards ahead the stream is

---

[74]*just at the head of the rapids*: A large camp, Amy's Rapids or Amy's Landing, was established at the head of a set of fierce rapids that stopped many gold seekers from traveling further with a boat. This camp was about four miles from the foot of Klutina Lake, and tents lined the river for a mile or more. It became the base camp for many men who were afraid they would lose their entire outfit to the rapids. Other men cached supplies here as they lined boats down the river. The KC used Amy's Landing as a base camp, which received its name from W. S. Amy. (PR)

[75]*everywhere may be seen goods drying*: Abercrombie, who had a tendency to exaggerate, wrote that about ninety-five percent of the gold seekers attempting to shoot the Klutina Rapids in the spring of 1898 wrecked their boats. (PR)

lost to view. The Horseshoe, Devil's Elbow, and Hell's Gate[76] are particularly dangerous and cause one to think before committing himself to their tender mercies. Lining or cordelling down is not without its danger and risk either, and quite a number have thus swamped their boats. With care, however, this is successfully done.

Two Germans were working their outfit down. One, fearful of the troubled waters, had walked on ahead to a point agreed upon where Lewey, who had nerved himself to run the rapids, was to land. Everything went well with Lewey until he attempted the landing. Here he struck a rock and upset. The German on the shore stood aghast for a moment, and then, throwing up his hands, he cried out to his struggling partner in the water, "Lewey, Lewey, you kin [can] have everything. I'm going home. I'm going home," and he immediately took the back trail. Lewey, as though he had not heard aright, yelled back, "Everything? Everything?" and grabbing a box as it drifted, he worked his way to shore. But these accidents are not comedies; they are tragedies. Every day sees the setting of the hopes of some poor soul, who with heavy heart takes the back trail. Indeed, the travel now is greater toward Valdes than toward the river.

The Devil's Elbow (Abercrombie 1900).

[76]*Devil's Elbow, and Hell's Gate* were approximately seven miles up the Klutina River from the CR and resembled the famous Hell Gate in the East River near New York City. Devil's Elbow is a mile further down stream. Abercrombie estimated the value of the lost supplies lining the river banks at Hell's Gate in August at $20,000. (PR)

Many are putting up permanent caches here and going on light. The reports from the front are not encouraging. Three miles above the head of the lake on the back trail, though, claims have been staked which pay 6¢ to the pan – $12 to $20 per day sluicing. Float gold is found on every bar on C.R., but not yet in paying quantities. Anything like thorough prospects will require a couple of months yet. Copper River itself is very rapid. Several of Co. K. have been down to it. By good fortune we got an opportunity to buy out a party 13 miles above the mouth of the Klutena R. These goods we bought at less than the ruling prices here. We then sold a like amount of our outfit here; and thus, while making money directly, we got ahead 800# of provisions 38 miles to the front, free.

The colonel got back from Valdes last evening. Richards and Egan came with him. They remained behind when we came from the head of the lake to bring the horse. The colonel found the surveyor had gone and Valdes dull. Our title will be all right. All we need is a strike, a big one on C.R. The glacier now is rather dangerous. Jackson, the mail carrier, lost a burro in crossing. He is going through with several animals this trip to Circle City,[77] another man taking his place as carrier. He is now a mile or two above here with his train. The Col. got from him all the mail he had for Co. K., and I was placed in receipt of your letter of March 26th and C.W.'s of March 24th. I was pleased to get them even though I had hoped for news of a later date. Our mail facilities are bad enough, aren't they? I hope my letters are being received more regularly. Deliveries and opportunities to send out mail will grow less frequent too, I believe, from this [point] on.

By the way I must speak here of Rayl. You will remember I told you that he had gained for himself the contempt of the boys. Personally, I was inclined to be lenient with him on account of his depressed condition. I guess the dislike was well merited, however. Returning, he and Henning had to stay about a week at Valdes before getting a boat. Their board-bill for this time, amounting to $34, they said would be paid by us; and Rayl, not satisfied with this deliberate misrepresentation, got Mr. Hall, who is in charge of our interests there, to advance him $3.50 on our account, saying he needed that amount for some purpose. I knew Rayl had the reputation of being up to sharp practices, but I did not think him dishonest. To further show the character of the man, in withdrawing he wanted the Col. to make him a better proposition than Henning, insinuating that Henning, as a relative of the colonel, had probably been made more special terms on becoming a member. Failing in this, he asked the Col. for a pocket knife for his boy. (We have a few for barter), saying that he had gotten nothing "out of the company." Poor fellow. The Colonel told him none of us had gotten anything "out of the company," – that we expected to get it out of the country – and refused him. I believe he said he would give Rayl $1.50 to buy a knife. For a wonder, Rayl did not accept.

In your letter you speak of expecting me home in the fall. To me now it looks very much like a two year proposition. Unless strikes are made on Copper River, and we are fortunate in locating claims, a number of our party will push on to the Tanana,[78] the Klondike diggings, or until something prevents our progress. I know how great a disappointment it will be, Father, to you,

[77]*Circle City*, Established in 1899 on the W bank of the Yukon River. (AP)

[78]*Tanana*: The Tanana River, central Alaska, tributary to the Yukon River. (AP)

to Mother, to Lizzie, and indeed to relatives and friends; but I have come too far and worked too hard to return so soon empty handed. It will be no less a disappointment to me, but I must feel sure of the "decree of fate" before taking the "back trail." If the good luck of Co. K. continues, there is no doubt as to its ultimate success. You all may rest assured that I shall take the best care possible of myself and that no unnecessary risk will be run. I will write again shortly, probably from Copper Center at the junction of the Klutena and the Copper, for we hope to begin lining down tomorrow. Write me all the news and enclose any clippings of interest. How did you come to sell the chickens? Tell C.W. I will write him soon. I am pleased to hear that Edith is getting along nicely; and thank her for her remembrance of me in her prayers, poor little thing. How is Snyder, and why have I not heard from him? How the pictures? By the way, have Al send another package of films, preferably those put in tin containers; for the mail carrier always falls in the river 3 or four times. I shall send out a roll with this. No late war news has been received.

Hoping that all are well, and with love,

Leroy

[On other side:]

Sunday, July 3, 98 – Our plans relative to running down the river were changed, and we are still in camp here. We deemed it advisable to do some prospecting to the east and west, so two parties went out Wednesday. They returned last evening with no finds to report. We shall now probably begin cordelling down the river Tuesday or Wednesday, probably the latter. Tomorrow, the Glowing 4th, will be celebrated by the camps along the trail. There will be a dance. A game of baseball will be played, boat races run, and shooting contests of all kinds. There is lots of patriotism here, and I hope for a pleasant day. For the past few days it has been cooler, and we have been troubled less with the mosquitoes.

Leroy

Letter 23 [No envelope]

Camp, Amy's Landing
Valdes Trail, Alaska.
July 29, 1898

Dear Father, Mother and Sister:–

I last wrote you July 4, and I hope you are in receipt of the favor by this time. Jackson passed through camp on the 20th, but I was again disappointed in receiving no mail. I have been worried lest my letters to you have miscarried, too; for I know how uneasy you would be. I have every assurance that this mail will reach you, though, as it will be taken out by Mr. Richards. He will leave for Valdes Monday. Sharp is going with him. Bailey went back on the 5th inst. He was going home. Clark accompanied him as far as Valdes and will winter there. This leaves six of us here: the Colonel, Bruce, Egan, Warren, Hiram Richards, and myself. Richards and I shall winter here or possibly at Copper Center, 25 miles below, as that is the largest camp and so afford to me the best opportunity for practice. As its name indicates, it is a central point and will be a base of operation.

The Colonel, Bruce, Egan, and Warren with three pack animals left Copper Center Wednesday the 27th for the headwaters of the Tanana, about 140 or 150 miles. They are traveling what is known as the Millard Trail,[79] a trail which cuts across the great bend of the Copper and saves about 50 miles. They will prospect the upper Tanana streams as thoroughly as their limited time and provisions will admit of and then pass on across the divide to Forty-mile,[80] if possible, and then to Circle City – perhaps Dawson.[81] It is not known whether this can be accomplished or not, but it is not the intention of the party to return unless something insurmountable presents itself or accidentally occurs. If any strike be made, or if for any reason it be desirable that we follow, the pack animals will be sent back; and we will cross. In case nothing is found, the course the Colonel will take is problematical. They are provisioned for only two months. But no strike being made, they expect to do packing and thus get supplies. By this time navigation on the Yukon will be closed; and if they do not deem it advisable to winter in the Klondike, they will probably take the Dalton Trail back to Dyea.[82]

I have now forgotten what I wrote you in my last letter of our plans at that time; but it was then intended that the Colonel, Mr. Richards, Bruce, Egan, Warren and Sharp would line a boat down the Klutena to Copper Center and then up the Copper River to the headwaters. On account of the great difficulty in getting down the Klutena, this plan was changed. About 4 miles below here the boat got away from the boys. Fortunately it was caught about 1 1/2 miles further on at Cox's landing,[83] where it ran into a small eddy. It happened that Sharp, who had gone on with the horse, had just reached Cox's ldg.; and he with several others were seated on the bank talking when the boat rounded the point into view. Sharp recognized her immediately; and as she struck the eddy, the boys sprang in and caught her. She was full of water and everything wet, but nothing had been lost. The other boys shortly came up, and camp was established. The next few days were spent in drying the provisions. Meantime the Colonel decided that he, with Richards and Egan, would take the horse and push on to Copper Center and then pack up the Copper. The other three boys – Bruce, Warren, and Sharp – were to pack sufficient to Copper Center to last 3 - 4 weeks and there wait word from the Colonel. At Copper Center the Col. succeeded in getting two more horses. Thursday the 14th inst., the Col. and party left C. Center. They crossed the Tazlina that evening. The next day they reached our cache above and increased their packs. I wrote you, you will recall, how we got the cache above the Tazlina. Friday night the party went into camp 7 miles beyond the river.

---

[79]*Millard Trail*: In the summer of 1898, a party of prospectors led by B. F. Millard, assisted by a party of former Great Northern Railroad employees, built this 70-mile trail, which began on the E bank of the CR near CC. It skirted the bases of Mt. Drum and Mt. Sanford and merged with the CR Trail near the mouth of the Slana River. (PR)

[80]*Forty-mile*: The Fortymile River, flows NE 60 miles into Canada as a tributary to the Yukon River. (AP)

[81]*Dawson*: City 1601 miles up the Yukon River from St. Michael on the Bering Sea, adjacent to "the creeks," the focus of the Klondike Gold Rush of 1897. (AP)

[82]*Dalton Trail back to Dyea*: The Dalton Trail, proceeded N by NW out of Pyramid Harbor in Chilkat Inlet (near present-day Haines) in SE Alaska, about 15 miles SW of Skagway and Dyea. The 350 miles was a practicable route for horses and livestock into the Klondike region. Dyea was a town at the head of Lynn Canal, first point of egress to the Chilkoot route to Lake Bennet, adjacent to Skagway at the head of Lynn Canal, existing from 1896 to 1902. Skagway and Dyea are accessible from the head of the Dalton Trail by boat only. (AP and PR)

[83]*Cox's Landing*: On the Klutina River, at or near the foot of The Gorge, a canyon a little below Klutina Lake. (AP)

Saturday morning brought disaster. As usual the boys built several smudges when they made camp Friday evening, the mosquitoes being terrible. During the night these burned down and apparently went out.

Saturday morning the horses had strayed, and the Colonel and Egan went in search. Richards, after putting the camp to rights, went off to cut trail. During their absence the wind arose, fanned one of the smudges into flame, set fire to the cache, and completely destroyed it. Richards came back just in time to save the tent and contents, while the Col. and Egan returned only to see the ruins of provisions, personal duffel, $80 in money, etc., etc. This was the worst hand Dame Fortune has dealt us. The boys now returned to Copper Center. The delay caused by the accident and the financial loss led Mr. Richards to decide on returning to the States.[84] Meanwhile Bruce, Warren, and Sharp at Cox's decided to run the Klutena from that point. To pack 50 lbs. 20 miles – the distance between Cox's and C. Center – is much like work. Having decided to run the river, Thursday the 14th, the boys sent for Hiram Richards and myself to help them. We immediately left for Cox's, and Friday noon we began the wildest ride I have ever experienced. As we started, Sharp was in the bow, Bruce and Richards at the oars, Warren passenger, while I took the stern. The first two miles we made without much trouble, though we shipped water and had to land and bail out. It happened that this brought us just a short distance above the Devil's Elbow. Sharp went on ahead to take a look at the river. In a short time he returned with a painful expression on his face and said he wasn't feeling well. He also said the river was bad. The rest of us then went forward to see if we could interpret what the "wild waves" were saying. They were holding a pretty animated conversation, and I think that all of us felt a little doubtful as to the result of " meddling" too much in their affairs. The Elbow is marked by a narrowing of the channel, shoal to the right, a big boulder in the center over which the water rushes to form a great swell below. To the left and a little beyond the rock mentioned, another juts from shore and turns the current somewhat abruptly to the center where swells 6 - 7 feet high mark underlying rocks again.

A short distance beyond, the river turns to the right. Warren and I decided on the course – to run to the left rock above close, swing to the right, and take the middle through the swell. We returned to the boat. Sharp said he'd walk. I then took the bow, Warren the stern, and we pushed off. We went through beautifully, though we shipped about 12 inches of water. I was standing in the bow when we took the first swell. As the boat's nose ran into it, though, the water came over full on my chest and seated me most suddenly. All were wet practically from head to foot. We went on several miles further and stopped for the night, "just escaped" [wrecking] being the case frequently. The next day we ran down to within a mile of Hell's Gate from which point we lined to the Gate and went into camp 10 p.m.

---

[84]*Mr. Richards to decide on returning to the States*: In an interview LST gave to the Beaver Falls *Weekly Tribune* (July 26, 1899), he says, "Hiram Richards showed signs of scurvy soon after it became epidemic and I had him go out to Valdes." It seems likely Townsend did not indicate this cause of departure in his letter in order to avoid alarming his family, principally his mother. (AP)

Above: Hotel and post office in early Copper Center. (Roppel-DeArmond Col, no #, Summer 1898) **Below**: Looking up the Klutina and at early settlement of Copper Center (Roppel-DeArmond Col., # 92, Summer 1898)

Saturday I did not get to bed until 3:30 a.m. Sunday we rested. Monday we lined through Hell's Gate and ran to Copper Center. From Hell's Gate to C.C. the trip was no less exciting, and "escapes" were frequent. Just as we landed, we struck a rock and stove a big hole in the boat. We made a successful run, though, and lost not a thing. Of course the cargo was more or less wet, though imperishable goods were placed on the bottom and all well covered with tarpaulins. We stayed at Copper Center over night; and after the arrival of the Col., Richards, and Egan Tuesday a.m., Hiram and I left for Amy's Landing, where we arrived Wednesday morning.

Since then we have been at camp here with only auction sales for excitement. The rush is on for Valdes, and everybody is selling out. Flour sells at $3 - $6 per C[85]; beans, 1¢ a lb.; Coffee, 5¢ lb. Clothes bring little or nothing, a $25 moleskin suit selling for $4.50. Fire arms sell for a song: a 30-30 Winchester with 1 or 2 hundred cartridges brings $2 - $3. Tools and hardware can almost be had for the asking. A whip saw sells for $25, a hand saw the same. Boats that would have brought $50 - $150 can be gotten for $1 - $5. The only things that command fair prices are the luxuries, so called: Sugar, 15 or 16¢; milk, 25¢ can; Cocoa 25 or 30¢ can; Eggs – Lamont's Crystallized, 75¢ - 90¢ lb. By the way, these eggs are a household necessity. With them you are never out of the equivalent of fresh eggs for cooking, for cakes, custards, etc. I question if they do not equal the fresh article. They make an excellent omelet, too. Mother need never worry about not having fresh eggs if she has Lamont's Crystallized. Try to get a sample in Pittsburg, Father, for her to try. If Lizzie is still going up, she can inquire for them. They cost $1 per lb., and this quantity represents 4 doz. eggs.

All this reaction is due to no strikes being made – (Newspaper accounts to the contrary notwithstanding). It is the history of all mining stampedes. By and by outfits will command nothing and will simply be abandoned. Already the back trail is marked by much discarded clothing, etc., particularly along the glacier, the trail across which is difficult and dangerous. As regards the likelihood of a strike this fall, I am in doubt. Colors are found everywhere, almost, in numbers varying from $2 - 30 or 40 to the pan. Some coarse gold has been found. But in almost every case it has been impossible to reach bedrock. At from 2 - 6 feet the water comes in and prevents digging further. Bar digging has been impossible, too, on account of the high water. Perhaps in the winter when all is frozen, bed-rock may be reached and pay dirt found. I am inclined to the belief, however, that bed-rock will be found only at such a great depth as makes any good finds unlikely. But Mr. Richards will soon reach B. Falls, and he will give you a detailed account of our trip and of our future plans. Sharp will probably call on you too.

You must not worry about me. We have all necessary supplies – provisions, clothes, etc. There will be no scarcity of anything. I have sufficient money to get out any time; and there will be some [medical] practice to bring in a little more. Game will be plentiful. Yesterday at the lake four miles above, a man was almost killed by a grizzly. The animal had been wounded several days ago; and the man, searching for a lost bowie knife, came upon him. The bear struck his head and face, tearing away the scalp and fracturing his jaw in two places. I have not seen him but understand he is in bad shape. He has been unconscious most of the time since the injury. He

[85]*per C*: per hundred pounds.

was armed but taken by surprise and had no chance to shoot.[86] This is the first accident of its kind, though bear are plentiful, and quite a number have been killed. They are now coming down to the river for salmon, the run of which has been going on for some time. This will interest Al.

I am still wondering how the pictures have proven, that is, such as have been exposed since leaving Valdes. Richards will take out a roll or two of films. I wonder if J.A. has sent me any more rolls. I hope I shall hear from you next mail and from Al. How is Mother, Sister, Grandmother, and all of you? Keep on writing, and in sending any news items, put in the form of "clippings" and send "First Class." The carrier will not pack papers, or at least brings but a few.

Hoping all are well, I am.

With love,
Leroy

Regards to all my friends.

Letter 24 [No envelope]

In Camp. Amy's Landing
Valdes Trail, Alaska
Monday, 10:50 a.m. **Oct. 10, 1898**

Dear Father, Mother, and Sister:–

I am afraid you will be worried on account of not hearing from me for so long a time, but this is the only opportunity I have had for writing since Mr. Richards left us Aug. 1st, carrying mail for you, which I hope you received. I trust he has long ere this reached B. Falls in safety and has given you assurance of our welfare. For a few days after he and Sharp left, we felt a little depressed; but we busied ourselves for a hunt with the Lewis party at the lake above. We wanted some fresh meat, and both Richards and I were anxious to get a shot at the big cinnamon bear that attacked and nearly killed the man there. We left camp on Monday following their departure and reached the lake about noon, tired of our heavy packs. The Lewis party advised giving up the proposed hunt for the cinnamon, as the great number of hunters there in search of him endangered one's life through accidental shooting. So we decided to go up St. Anne's River, ten miles up the lake.[87] Accordingly we spent Tuesday in making preparations, Wednesday morning left Lewis' camp in his two Acme Canvas boats. and started to row up the lake. We had gone about eight miles when the lake became so rough we had to put ashore and camp for the night. Thursday morning we broke camp, rowed to the river, and lined up about 3 miles where we made a permanent camp. The same afternoon Mr. Lewis and I went hunting. But though sign was plentiful, we saw no bear. Every bar in the river was covered with tracks where bears had come for salmon: big tracks, little tracks, all kinds of tracks. One was 7¾ inches across. Friday morning

---

[86] *grizzly bear*: Remington tells how the 23-year old encountered a bear on one of the salmon streams. The bear attacked, tearing the man's scalp loose and then broke his jaw (he says in five places). The man managed to return to his boat and lift the sail. Near Klutina City camp, he was rescued and taken to Dr. Ottaway's tent. Later Townsend and Ottaway performed surgery, removing shattered bones and then plastering his jaw up in blue clay taken from the bottom of the lake where it emptied into the river. The young man was helped to Valdez and returned to the states. (PR)

[87] *St. Anne's River ... lake*: This "creek" and lake, according to Remington, was named in honor of Mrs. Anne Barrett, who is ever-present in his account of the summer of 1898 in the CR valley. (PR)

we spent about camp fishing. We caught many trout and grayling. One would rise to the fly almost every time it was cast. My! what fine fish and fishing, and how we enjoyed the fries.

Friday afternoon Mr. Lewis and I packed our blankets about 8 miles up river, where we took stands and watched for game. None came into view, however; though a bear passed through the timber a short distance back of us. We could hear him plainly; but he offered no shot; and we returned to camp in the morning empty-handed, in time for breakfast after three hours travel. The balance of the day we spent in sleeping and fishing. A splendid string of fish was taken out of the wet. Saturday evening I again took a stand and watched for bear but was disappointed, though I was out until 10 p.m.; and the boys, all of whom had returned, were becoming uneasy.

Sunday morning we broke camp; lined down St. Anne's, and took our course toward Lewis' camp, which we reached about 3 p.m. The ride down was delightful. The day was bright, the sky cloudless, and just cool enough to make rowing pleasant, – a day such as Indian Summer brings at home. About 1 o'clock, as we rowed quietly along far from shore, came the notes of a bugle. Instantly every oar stopped, and only the lapping of the waves and the quicker beating of our hearts disturbed the strain of "Marching through Georgia." We thought of the "Brave Boys in Blue" at home and wondered how they fared; and as the last notes echoed and reechoed from the mountains, we gave the bugler a rousing cheer. He responded with "Yankee Doodle," and again we cheered to the echo. Then came the notes of "Home Sweet Home," sweet and low; and thoughts went to loved ones at home. I took my place at the dinner table with you all – Mother on the right, Father to the left, and Sister and Lou opposite. I ate my fill of roast beef, of potatoes, of fruit, and of sugar, and praised mother's cooking. But the sweet notes, echoing and reechoing fainter and fainter, gradually lost themselves in mountain and valley and brought us back to realities. Rowing in silence, no one caring to catch another's eyes, finally again we cheered the lone musician and rode on in silence.

Opposite page: Of the four persons in the photograph, two are known. L to R: Mr. Tom Temple and Leroy Townsend. The third person may be Mr. Washburn of the Lewis party. A fourth person resides outside the picture. Believed to be taken at Amy's Landing. (Snowden Col., September 1898)

Monday we prepared for a hunt in another direction; and about noon four of us – Richards, Lewis, Cooley, and myself, with four days' provisions and our camp equipage, bedding, etc., packed to a lake about 6 hours distant and went into camp. We hunted hard until Thursday but saw no big game; no fresh sign was seen indeed. We did get several grouse and a number of pine-squirrel. Accordingly we returned to Lewis' [camp] Thursday evening. We expected to return to our own camp Friday, but a miners' meeting was to be held that same evening at which Capt. Abercrombie[88] was to speak, so we remained until Saturday. Abercrombie addressed the men relative to their powers in establishing rules and regulations for their own government, etc.

Saturday we returned home. We were accompanied by Mr. Tom Temple, a colored man and the practical miner of the Lewis party, as our guest. We entertained him until the Thursday or

[88]*Capt. Abercrombie*: Captain W. R. Abercrombie, Second United States Infantry, Commanding Officer of the Alaska Exploration and Relief Expedition. (AP)

Friday following, at which time Mr. Washburn, another member of the Lewis party, called and told us that it was deemed advisable to make a trip to the Tazlina River for the purpose of establishing a mining district and taking claims. Mr. Temple had just returned from this stream and had found good "prospects," and later news from friends there decided the boys to go over.

Mr. Temple accordingly returned to the lake, and we began preparations for leaving. It was our intention, too, to build a cabin on Hudson Lake,[89] as a party of us, numbering eight had about decided to move in this direction this winter. This is along the course of the old Capt. West Trail and necessitates our pulling on the back trail about 15 miles, then 15 miles up St. Anne's River[90] to the lake, which is about 10 miles long. From this point we shall move our outfits at the earliest opportunity to Tazlina Lake in which the Tazlina River has its source and thence proceed toward the Sushitna[91] River. From Hudson Lake to Tazlina is about 12 miles, etc. Thus you will see we have much hard work before us this winter.

Well, we left camp here Sunday, Aug. 29. Our trip was to take us about two weeks. We reached Lewis's Sunday evening, tired of our heavy packs. Monday was spent in further preparations there; and Tuesday eight of us set out, our party being as follows: Lewis, Temple, and Washburn = the Lewis Party; Mr. Cooley, a young man from Florida (who, by the way, was well acquainted with poor Joe Taylor who worked for you so many years. He did not know of Joe's death); Mr. Hosford, a practical western miner; Mr. Buer; [Hiram] Richards and myself.

Four of us, two in each of Lewis' two canvas boats, rowed up the lake to the mouth of St. Anne's River, a distance of 10 miles. The others took the trail and met us there. We then began roping up the stream. This was hard work. The stream is deep and shallow by turns, grown over with brush and pines; it is so crooked that three miles of travel means only 1 as the crow flies, and the water is icy cold. For four days we waded and pulled before we reached Lake Hudson – distant from the mouth of St. Anne's about 13 miles. We were wet almost constantly; and the second day and night it rained so that we slept in blankets as wet as our clothes; and our camp on the sand bar, called Camp Grayling on account of the large number of such fish caught, was hardly delightful. But the morning of the fourth day brought us to the lake; and we enjoyed a pleasant row to a point about 5 miles up the right shore, where we made camp, Moose-horn Camp. We saw hundreds of duck, later thousands – the lake being a feeding place. A number were killed; and we enjoyed duck stew, a pleasant change from the fish we had been having. While at this camp Richards saw a bear but unfortunately did not get a shot. Another passed within a short distance of our camp one night. One of the boys heard him, and we found his track next morning. High winds kept us at this camp 3 days. It required little rowing, for the wind was blowing a gale and the waves were running high; but it necessitated good management of the little canvas craft to keep from swamping. We quickly threw up a windbreak, for the wind was chilling, and thus was established Camp Mallard – the name being suggested through the number of such duck seen and killed, but particularly on account of my killing two of the finest specimens at one shot with a rifle – both shot through the head.

---

[89]*Hudson Lake*: Comparing Dr. Townsend's description with today's USGS map, this would be today's St. Anne Lake. At some time the name changed. Remington reports the lake was named for Professor Hudson of the Smithsonian Institution. Hudson, he states, was doing scientific work off and on in the vicinity during the summer of 1898. The USGS map still shows a trail to St. Anne Lake originating on the north shore of Lake Klutina. (PR)

[90]*St. Anne's River*: Actually termed St. Anne Creek, leading north from Klutina Lake to St. Anne Lake. (AP)

[91]*Sushitna River*: Early Native name for the Susitna R, tributary from Cook Inlet to the N. (AP)

We remained at this camp several days, building caches, locating cabin site, etc. We then made up our packs for the trip to Tazlina about 10 miles distant. They were heavy, weighing from 60 - 80 pounds as we were carrying extra supplies so that two of the party, Temple and Hosford, might continue on a 3 week's trip, prospecting the country from Tazlina Lake through the Sushitna River. The trip from Hudson Lake to the Tazlina R. usually requires but one day. Unfortunately we decided to "cut across country" instead of following the trail. The result was far from satisfactory. We soon saw our mistake, but our search for the trail was fruitless. For 4 days we packed through timber and swamp before the Tazlina was before us. We worked our way through the thickest of timber and underbrush; we picked our way around or through miles of swamp and here and there skirted lakes of various sizes. We camped in bog and were thankful frequently for the brackish water we found there. We wended our way over bluffs like the Alum Rocks[92] where a slip or misstep would have been unpleasant. Fortunately we had plenty of provision; and on the afternoon of the fourth day, after packing through a drizzling rain, we reached the river. It is astonishing how easily one loses his bearings in dense timber when he has not the sun to guide him. At one time, neglecting to look at our compass for just 15 minutes, we found that we were traveling due south when our course was directly north. We had completed half the circle. After that our compass was never out of our hands. Many a poor prospector who has started alone in his search for color alone will never be heard of, and not a few "parties" are reported "unaccounted for." A Polish physician is reported "lost" in this same country we crossed.

Upon reaching the river, we made camp on a bar, got our supper and turned in. Before retiring we dried ourselves pretty well and so got a good night's rest. The next day it rained and we did not break camp. The morning of the third, though, we made an early start up the river and reached Miller's Cabin, our objective point, the same evening. Miller had been working a bar on the river some and was taking out some gold, though not in paying quantity. Two men took out $5 in 6 hours with a rocker on account of the high water. We spent several days with the boys here, panning and prospecting about, but we found nothing worth "locating." A district was formed, however, in anticipation of the freeze up permitting us to accomplish more. I presided at the meeting. On account of being short of provisions, our party now divided. Four returned to Lake Hudson.

Washburn and I, with Temple and Hosford, went up river to Tazlina Lake in order to pack sufficient "grub" for T and H to continue prospecting from that place. Miller had joined the party, too; and for the first 5 miles he had two of his "partners" give us a hand. To the lake was about 10 miles, and the second half of the trip we all carried packs weighing not less than 80 pounds. Around clay-banks hundreds of feet high and in which we had to cut foot holds to pass, through the quick-sand, up and down the sides of deep ravines, and through the thickest timber, we wended our weary way. We camped one night on the trail, reaching the lake about noon the second day. The tent and few supplies which Temple and Hosford had left on an earlier trip we found all right. But the horse had wandered up the lake and not been found when we left the next morning for Hudson, though the boys had tracked him 6 - 8 miles. Washburn and I reached Lake

[92]*Alum Rocks*: Rock cliffs made of alum across the Beaver River from New Brighton, Pennsylvania. (PJT)

Hudson in the evening about 8 o'clock, just in time to join the boys at supper. We were hungry, though we had fared better during the day than expected, as I had killed 3 grouse, two with revolver and one with rifle. These we stewed for dinner in an old tea can which we found, hobo-fashion. Tea we made in an old corned beef can. I wish you could have seen us as we prepared that meal. It is wonderful how one can manage. But we enjoyed our dinner. We had brought a little bread with us, but necessity demanded that we leave all possible provision with the boys behind. We enjoyed a good night's rest at our old Camp Mallard. The next two weeks were spent in making hay for Lewis' horse and in building a cabin. During this time we had all the duck and grouse we desired. There were thousands of the former on the lake, congregated there prior to their migration.

Wednesday the 28th we started for our home camp. We reached Lewis' camp the same evening, having come about 35 miles, all by water. But the wind was in our favor on both the lakes, and we shot St. Anne's River in the canvas boats as we were "light." Washburn and I brought down one, and Richards and Cooley the other. Both boats came through without being snagged, and in this the trip was remarkable; for as I have before stated, the stream is swift, crooked, and overhung, to say nothing of snags and rocks. I regard the Acme Canvas Canoe as a most excellent craft. You may realize that the run was no child's play when I tell you that through exertion and nervous strain we were all constantly in a drip of perspiration. It did not equal running the "Laughing Klutena," though. Richards and I returned to our camp on Thursday. Since then we have packed one load up from Cox's – mine weighing 88 pounds, a distance of 6 miles, and one from Copper Center, 25 or 30 miles. The round trip of the latter took us 3 days, but the return with 50 - 60 pounds we made between 9 a.m. and 9 p.m. Such a trip is too much, though, and we were all more or less "played out," = footsore and weary. I escaped the first, but my knees were pretty sore, and I must plead guilty to the second. Think of packing a sack of flour to Pittsburg for a day's work – delightful (?), isn't it?

Of late there have been a number of "excitements." A find of some coarse gold on Quartz Creek, a tributary of the Tonsina east of here, caused a stampede in that direction.[93] Miners left at all hours of the night. One party which was being followed by a second, in order to throw the second off the track, threw down their packs as they neared the "find" and belted in every direction. That night they went without food or bedding, but the next day they made locations and returned to hunt for their duffel. Bear Creek, a small stream emptying into Quartz, has been staked too. The Teikel[94] has produced an excitement; and Fall Creek,[95] one of the tributaries, is all taken up. There is great difference of opinion among the men as to the value of these claims. In general they are not thought much of, and the best claims can be bought for a few dollars. $100 is the

---

[93]*Quartz Creek ... a stampede*: Some placer gold was discovered in the gravel of this tributary of the Tonsina River. However, the placer deposits were mined out in 1898-99. The U. S. Geological Survey estimated the production of the creek probably did not exceed $1,200. (PR)

[94]*The Teikel*: A tributary of the CR, S of CC. (PR)

[95]*Fall Creek*: Diggings began about a mile above the creek's confluence with the Kanata River and extended five or six miles up the creek. The gravels, of glacial origin, were from a few feet to 600 feet in thickness, but coarse gold nuggets, valued at $5 or $6, could be found where bedrock was reached. This area was not an important producer of gold in later years. (PR)

highest yet paid, and only one has brought that figure. Many do not expect to return to their claims, thinking them worthless. Fifty-cents is the best that any one pan has produced, and this was gotten through crevicing on bed rock. A number of coarse pieces worth from 10 - 50 cents have been found, however. I have seen one or two. Some think they are just on the margin of a pay zone. A reported find in the direction of Mount Drum[96] has caused quite a number to "go" in that direction, and Wetstone Creek has been located for about 10 miles. This creek has its source in the vicinity of Drum and empties into Copper River about 15 miles below Copper Center. In general our opinion of the above is that they are "no good," though one or two of the claims on Quartz Creek may be worked to a profit.

This embraces the "news" of camp since Mr. Richards left to the present time. You will notice that our trip to the Tazlina took us much longer than we expected; we were gone five weeks instead of two. During this time I had my clothes off but once, when after a little flurry of snow I took a bath.

A worn Leroy Townsend appears right, at base Camp, Amy's Landing. (Snowden Col., September 1898)

[96]*Mount Drum*: Northeast of CC. (AP)

Toward the last of our trip it became very cold nights. Snow had covered the tops of the mountains, and a little had fallen in the valleys, though not to stay. I am now writing Sunday p.m. at 2 o'clock, Oct. 16 – 98. About an inch of snow has fallen, and it still continues snowing. The thermometer registers 28°, a little warmer than it has been for days past; and the quiet with the snow gently falling makes it seem like a holiday. The nights of late have been very cold, 10° and 12° below 0° having been registered. But one does not feel the cold as we do at home, and I have only started wearing my coat, and across the way the Indian children are wading in the icy water knee deep. I saw them this morning. They have been bare-foot, too, most of the time, though I presume with this fall of snow they will don their moccasins.

Monday a.m. This morning the Indian children have celebrated the first snow by taking a swim in the Klutena. They came down naked and were in the water 5 or 10 minutes.

Five Native children of the Klutina and Copper rivers region (wearing moccasins). (Snowden Col., Townsend, circa October 1898)

We have done some trading with this small band across the river. They are under Chief Stickman[97] and made camp there a week or ten days ago. We sold them some few articles, traded for a beaver skin and one or two horn spoons as made by them from the horn of mountain goats. One I got from the Chief himself is particularly large and symmetrical. I hope I shall be able to get it back to B. Falls. The Indians are honest and perfectly harmless. They are quite intelligent and well up in trading. But they are a fast-dying race and will soon be of the past. The gaunt specter of starvation enters too frequently their teepees. The government should assist them.

"Chief Stickwan (R) and Comrade, September 15, 1899" (Caption Captain Ambercrombie's) (Abercrombie 1900)

[97] *Chief Stickman* (variations are "Stickwan" and "Stephan") headed a small group of CR Indians. Their territory encompassed the area between the Klutina and Copper rivers south of CC. (PR)

Above: Natives on the river, a woman (aft) waving to Townsend. **Below**: Two Natives before an abode (camp/sweat lodge), both smiling, one broadly, for the photographer. They are either just in or preparing to go out: clothing, snow shoes, cooking utensils, pack, and rifle are spread before them. (Snowden Col., Townsend, circa October 1898)

Ahtna on the trail, well equipped and assisted by a dog with pack. (Snowden Col., Townsend, circa October 1898)

☆ ☆ ☆ ☆ ☆

Sales go merrily on. There are two taking place today and one announced for tomorrow. Probably others will occur during the week. Prices are ranging higher. Flour brings $12 per C., beans 4¢, sugar 25 or 30¢, baking powder $1.00 lb. Foot-wear is high, but clothing in general very cheap. There will be no suffering here this winter due to lack of provisions. There is enough to supply several times the number who will remain. If reports to the contrary appear in the papers, do not give them any credence. I am well prepared to meet Alaska's winter blasts – have plenty of clothing, and we are well supplied with provisions. Mother asked regarding washing and mending. Yes, we do both ourselves, but she must not worry. I don't do very much. Ha, Ha. We do our own barbering, too, what little is done. I had my hair cut a couple of weeks ago, and it is now shorter than it has been for years. It is grayer, too; for the silver threads are fast increasing, a peculiar feature of Alaska; for many have noticed this same bleaching.

Being barbered in a camp-built barber chair, Amy's Landing. (Snowden Col., circa October 1898)

☆ ☆ ☆ ☆ ☆

We shall start within a few days to put up a cabin here. Our logs are all cut now. We deem this advisable even though we move shortly toward Hudson Lake and there put up a second. Had we not been so long on our last trip, it would have been finished long ago. Cabins have gone up everywhere, though many expect to winter in their tents. Above us are two of the finest on the trail. The boys burned brick and built a big open fireplace and oven. The interior is elegant.

An "elegant" interior. Leroy Townsend is second from left. The prospector to his right struggles to hold the cabin's big black kitty for this communal picture. Kitty was apparently not patient as the prospector's hand is blurred (moving) in maintaining a grip on kitty's scruff. (Snowden Col., circa November-December 1898)

☆ ☆ ☆ ☆ ☆

Along the trail is no sickness. Alaska is a wonderfully healthy country. The summer in this valley of the Copper is delightful; the only drawback: mosquitoes. I anticipate nothing like the severe winter we heard of in the States. I expect very cold weather and storms and blizzards, but according to the Indians the snow fall is not more than 2 feet here. This is probably correct; careful observation reveals no moose-yards,[98] which we might expect if the fall were very great.

X X X X X X

Under separate cover I shall send two rolls of films to Snyder. I hope they prove good. I was greatly disappointed that the exposures taken over the glacier were spoiled. I could hardly have failed to get something out of 24 snaps. Many of the boys overexposed their pictures, but most of them got something. I was careful to consider light, stop, and time of exposure. When I return I shall try to make exchanges with those who were more fortunate in getting pictures of the glacier.

Above: Currants and/or berries aplenty. Two women and three men, several bearing the bounty of a harvest picked in the Copper River Valley. **Opposite page**: " currant duff, taken at our camp on St. Anne's" Lake. The playfulness portrayed in this photograph underlines pioneer needs and values: tremendous appetites and a special treat, with the six-gun symbolizing how each man will have his share and no more. (Snowden Col., Townsend, circa September 1898)

Did the pictures of the currant duff,[99] taken at our camp on St. Anne's, prove good? Currants were very plentiful here in season; and we enjoyed them greatly, making many currant duffs. I

---

[98]*moose-yards*: Moose yard, an area in which the snow is trodden down by moose, where they remain together during the winter months. (AP)

[99]*currant duff*: Currants in a duff, a flour pudding boiled in a bag, a dumpling. (AP)

never saw any cultivated varieties at home finer than the wild kind here. The only drawback was the scarcity of sugar. Had we had plenty of the latter, we would have put up some for the winter.

☆ ☆ ☆ ☆ ☆

And now I must make answer to some of your letters, six of which were awaiting me when I returned from our trip to the Tazlina. All the regular and one or two special carriers to whom I had given our names had mail for us. I presume they had made a division, as they have been blamed with doing before, in order to increase their receipt. But I never paid out any money with better grace than I did the $4.50 charged by the several postmen. I received the package of films from J.A., and one of the six letters was from C.W.

In the letter of May 22, you speak of Mother and you just having finished your walk around the farm. I have often thought of you doing just that and wondered how everything looked. I have sat with you all on the porch just after a cooling shower, as we used to. I have been at the spring and have looked at the chicken coop, now deserted. I have been at home on Sundays. I have wondered whether Mother still tires herself out or whether she worries about me. All this and more. I have asked myself if all were well and thought of each one individually. Urge Mother to take it easy as possible and to have no fears for me. Let her have confidence in me and be assured that all will be well this winter. There are only a few of us left; but we have the courage and the determination; and who knows but what with the coming of spring, Dame Fortune may smile upon us. I am glad to know Lizzie was still continuing her music and making progress. I hope she has kept it up and that she is well. I think of Grandmother, too, and of Aunt Lou.

Under separate cover I am writing to Mother in answer to her letters enclosed with yours. I dream of her and of home and again and again build air castles for the future and hope on.

I am well and cheerful; and the hearty way in which my many friends say, "Hello, Doc, you're gettin' fat, ain't you?" leads me to think that at least I am holding my own.

I have a tooth or two the fillings of which are "leaking." I expect to have them tended to when next I go to the Lake. There is a dentist there.

It will please her to know that I am held in high esteem along the trail. I think I may claim this without egotism.

I deem it wise in Mr. Bannon to have given up his trip to the Klondike. If the difficulties to be overcome in reaching there are like those presenting on the Valdes Trail, the hardship is too great for a man of his age. If the Leigh party met with equal difficulties, I predict an early return of a number of them. We have had two suicides on the trail, and quite a number have become mentally unbalanced.

☆ ☆ ☆ ☆ ☆

I wonder if Mrs. Tatro received my letter expressing my sorrow at the death of Mr. Tatro? I wrote about the 1st of June.

☆ ☆ ☆ ☆ ☆

The Rennick party we have lost track of completely. Though the country between here and the mouth of the Copper has been prospected on both sides of the river, we have received no news of them whatever.

☆ ☆ ☆ ☆ ☆

In the letter dated Aug. 12, Lizzie says, "Bishness was yust boomin." I should say it was. Has it all come to pass? The bridge built, the trees cut down – what trees? I don't know of any we could spare. George gone to Denver, Mills and Guppy gone, Joe Close at work, etc., etc. Well, it do beat all.

☆ ☆ ☆ ☆ ☆

I regret to learn of Aunt Hannah's[100] illness and hope she is well again.

X X X X X X

Grandmother and Little Edith,[101] too, I hope are better; indeed I am anxious for the health and happiness of all. Poor Little Edith. I wish she might grow strong and rugged. To think of her and then to see these little Indian children 3 or 4 years old paddling barefoot in the ice cold water in seeming enjoyment. It is all so strange, so very strange.

X X X X X X X

Tell Aunt Lou to possess herself with patience. All will be well.

X X X X X X

Leanna[102] encloses a letter with that of C.W.'s, which I was pleased indeed to receive. I trust indeed that Grandmother and Little Edith are well again. I wish, too, that I might enjoy some of the products of the garden which may have been preserved for the winter.

X X X X X X X

In his letter, C.W. speaks of the return of Rayl June 29. Have either you or C.W. talked with him lately? The meeting of Mr. Richards and Rayl has no doubt resulted in an interesting conversation. Is his conduct generally known to the people of the Falls?

C.W. tells me he [Rev. Richards] had the "grip," too. I hope he is enjoying good health now. I have thought of him frequently. You will, of course, show or read to him this letter, which must take the place of a personal one.

X X X X X X X

Present my regards to J.A. and wife and to folks at the Globe,[103] and in general to my friends.

How many times I have thought, "If they could only see me now."

Friday, Oct. 21, 1898

It is barely possible that the plans outlined above may not be followed out.

I have gotten some advice which leads me to think better of the Teikel "excitement" than I did at first, and I shall probably go over and investigate. It is too late to prospect very satisfactorily, but we shall try and reach bed rock by spring.

☆ ☆ ☆ ☆ ☆

I will try to get a letter out next month, but this may be the last to reach you until spring. Should it be, do not worry; and, indeed, do not become alarmed if you do not hear from me in the early spring, as our movements are somewhat problematical; and we may be too far from the post. Be assured of a letter if we can possibly get it out. We understand the *Valencia* will make a trip next month.

With love to all,
Leroy

---

[100] *Aunt Hannah*: Leroy's aunt. (PJT)

[101] *Little Edith*: Daughter of C. W. Townsend. (PJT)

[102] *Leanna*: Mrs. C. W. Townsend, Leroy's aunt. (PJT)

[103] *Globe*: Beaver Falls had several newspapers so titled, which had merged with other newspapers, such as the *Beaver Star*, to which one of these is probably referred. (PJT)

Letter 25 Postmarked JUNEAU, ALASKA, JAN 24, 1899; [on reverse] Chicago, Feb. 9, 1899; received Beaver Falls Feb. 10, 1899

In Camp. Amy's Landing
Valdes Trail, Alaska
Monday, 10:50 a.m. **Oct. 10, 1898**

Dear Mother:–

I have been writing to you all at home, but I felt that I must write a few lines to you alone in answer to your loving words enclosed in Father's and Sister's letters. Have I been remiss in neglecting to do so before? Then I must ask forgiveness. It is not that I have been forgetful of you nor of home, and you must never again reproach yourself for any lack of kindness to me. The remembrance of your love, and of Father's kindness and of Sister's affection are always before me; and I am going on, even with the hope that "Fortune" may smile and that I shall then return to show my appreciation of that Love and Kindness, and Affection. Do not worry about me; and as the weather comes on, think of me as surrounded by all the essentials of comfort, if not the luxuries: plenty of food, good clothes, warm bedding, and a cheery cabin, with substantial, whole-souled companions. Be assured that I shall run no unnecessary risks and that I shall take the best possible care of myself in every way. In thought, too, I shall frequently be at home with you around the fire, and together we shall read the periodicals and enjoy a quiet time. Take care and do not overwork yourself, and get as much rest as possible.

With love, Your Son,
Leroy.

Letter 26 Postmarked JUNEAU, ALASKA, JAN 24, 1899; [on reverse] Chicago, Feb. 9, 1899; received Beaver Falls Feb. 10, 1899

Camp, Amy's Landing,
Valdes Trail, Alaska.
Wednesday, Nov. 9, 1898

Dear Father, Mother, and Sister: –

An opportunity offers to send out letters, and I take advantage to send a few lines. I hope you will receive the letter I sent out three weeks ago. In it I went into detail regarding our proposed move toward the Sushitna River. I also spoke of the several excitements which had taken place; and while expressing the opinion that there was nothing in them, I added that later advices led me to think the Tiekel River proposition[104] worthy of investigation. I have just returned from there and am pleased to say that it is the best appearing country that I have yet seen in Alaska. I was fortunate enough, too, to make locations for Richards and myself, a claim of 20 acres in Richards' own name and an equivalent claim for myself as one of a party of eight who took up 160 acres. Of course the value of the property is wholly problematical, but its appearance leads me to be hopeful. Good prospects have been found, but no one has yet gone to bed-rock. This

[104]*Tiekel River proposition*: The claim addressed here was apparently filed. Townsend eventually trades a half share in it with W. J. Woliben, a volunteer in the CC Hospital, for developing it (see Appendix E). (AP)

has been impossible on account of the find being so late in the season and the distance so great, 60 miles over a most difficult and not un-dangerous trail. The trail leads over two divides, one of which is very high – always above timber and no water. So late in the season here a heavy fall of snow may be expected at any time, and such a storm might prove fatal to a party caught in it.

We were more or less apprehensive all the time we were out. Fortunately we had but a light fall of snow; enough, however, to obscure the trail in places and prove very annoying. We were out 15 days though we expected to be gone not over 10. Four were in the party. We carried 30 lbs of provision each, which with bedding, etc., made packs weighing 70 lbs or more, a terribly heavy load to pack over such country. We were 7 days in reaching the river, but one was lost through the appearance of stormy weather for crossing the summit. We traveled from 5 a.m. to 4:30 p.m. and without dinner in crossing this divide. Even then we did not reach timber on the other side. 4:30, at which time it becomes dark now, found us in a canyon, too, with almost perpendicular sides and little wood-drift. We had to camp. Actually it was impossible to find a place for our sleeping bags, the hillsides were so steep. I did not get any sleep, and during the night it snowed about an inch. One of the boys had to get up during the night and hold the sleeping bag of another while he crawled out. He was just about to drop over a 6 ft. embankment. Hardly any night did I sleep more than till 12 o'clock it was so cold. The last five days of the trip were on "rations," too, which was unsatisfactory; for we had only two meals daily anyhow. But I've gotten filled up again now, and everything is lovely.

X X X X X

When I reached camp I found your letter of Aug. 26 awaiting me. I much regret to learn of the suicide of Martin Metzger and of the death of Mr. Myers. Don't a few short months work wonderful[105] changes?

Has J. J. Paff gotten things arranged OK? I trust he will be able to do so.

With regard to the cook and his leaving, I never looked upon it as a discharge. He returned last July, but his leaving could hardly be considered a discharge. He was dissatisfied, and the Col. thought well of his withdrawal. He and the Col. did not pull together, and indeed none of us felt sorry to see him go.

I am delighted to know that the pictures are proving O.K. now and hope J.A. has received more films. I have sent several rolls and hope you will duly receive them with the letters forwarded.

I saw a bear this last trip but did not get a shot. Tell Mother not to worry herself. We have much here to enjoy. I do wish, though, that I had some of the corn she speaks of. I hope you are all well.

Mr. Richards has arrived in B. Falls, no doubt. My regards to him and to all the folks.

With love,

Leroy –

Will write at length at the first opportunity.

L

---

[105]*Wonderful*: In the senses of *strange, astonishing*. (AP)

Thanksgiving dinner announcement, read by a heavily bearded Leroy Townsend. "*THANKSGIVING DINNER* / NOV. 24TH / LIMITED NUMBER OF TICKETS – APPLY AT — CAMP HAAS / MUSIC / FIVE P.M." The other notice on the plank is not readable. (Snowden Col., November 1898)

Letter 27[106] Postmarked ORCA, ALASKA, APR 24, 1899; [on reverse] received Beaver Falls May 5, 1899

Copper Center, Alaska
April 3d, 1899

Dear Father, Mother, and Sister:–

For days now we have been expecting the mail, but it has not arrived. A late arrival from Valdes reports no boat in up to the 29th of Mar. Thus we are still in doubt as to what the government may do for us here. The conditions existing, however, are greatly changed – marked improvement having taken place. Discharges from the hospital service have been so numerous that only 9 patients are now in my care. Six are leaving for Valdes today. The reports from up the Copper River are much more encouraging too. One death has occurred since I last wrote you, but I now

[106]*Letter 27*: 145 days have passed since Letter 26 (dated November 9, 1898). The tenor of Letter 27 reveals that apparently at least one, and perhaps more, letter was written, detailing conditions in CC, for it presumes previously conveyed information. Additionally, in this letter Townsend indicates a previous letter. It seems likely that correspondence and film sent home during this period was never delivered. (AP)

feel justified in predicting no more, if we are at all fortunate in getting supplies in. Our volunteers, after a hard trip of 23 days, brought over 100# potatoes and about 40# of onions. To this we have since added 30# potatoes and onions and 6 dz. lemons by purchase. This is the available supply at Valdes until the arrival of another boat.[107]

Potatoes and onions sold at 50¢ per lb., lemons 25¢ each, etc. The camp now only numbers about 16, 9 of them sick.

Terrible experiences have been had on the glacier by parties crossing. On Feb. 26th, 7 perished, and on and about the same date others narrowly escaped. Six bodies have been recovered. The body of the 7th victim, Dr. Logan, has not been found.[108] Dr. Logan has been giving service at 12 mile camp, 4 miles this side of the foot of the glacier, all winter; and he with 4 companions were sledding out two scurvy patients when they were caught in a storm and perished. Richards crossed March 20th in safety.

I hope I shall be in receipt of many letters and clippings from you in the next mail. I am anxious to get word regarding the company's prospects and the movements of the Col. If they are at all encouraging, I shall try and cross to Forty Mile. Capt. Abercrombie is expected at Valdes anytime now, and upon his arrival I hope to learn regarding our chances of remuneration from the government for our hospital service.[109]

With love to Mother, Sister, and yourself,

Leroy

Letter 28[110] Postmarked ORCA, ALASKA, APR 24, 1899[111]; [on reverse] received Beaver Falls May 5, 1899

Copper Center, Alaska
April 13th, 1899

In my last letter I wrote you that we were expecting the mail anytime. It reached here Monday last, and I received your favors of Feb. with enclosures from C.W. and Edith, together with

[107]*available supply*: It is unknown who provided these supplies; Brown reported he sent "some green potatoes and sauerkraut." (PR)

[108]*Dr. Logan, has not been found*: Remington reported that Logan fell into a crevasse and his fellow prospectors tied their pack ropes together and let one of the party down, as he cut steps in the ice. However, the rope was not long enough and the doctor slipped out of sight. Some of the men went for more rope, which arrived in a few hours. By then Logan was almost overcome with the cold, but asked for a "last good-bye to John Barley Corn," which someone gave him. His body was not recovered. Remington's account does not appear in the Coroner's Inquest (Appendix D). (PR)

[109]*claim against the government ... hospital service*: this claim must have been referred to in undelivered correspondence. (AP)

[110]*Letter 28* is hand addressed as follows: "L. G. Townsend, Beaver Falls, Penna, Agt. Penna. Co. Agt. Penna. Co.," whereas earlier letters were sent in pre-printed stamped envelopes. Since on this postmark *Apr* is stamped upside down, as was the case for Letter 27, I assume it was also sent from Orca. It shows being received in Beaver Falls on May 5, 1899, as was Letter 28. There is obviously a letter missing, at least one, recording the move to CC, the illness and deaths of a number of people, and Leroy's work there as doctor, commencing officially on January 31, 1899 (Appendix C). Since Leroy's letters were tied together and have "30" as the total number recorded on the first envelope and there are still 30 letters in the packet, it would appear that any letters sent between the present numbers 25 and 26 and before 27 never reached Beaver Falls. (PJT)

[111]*Apr* is stamped upside down but not the date and year. (PJT)

letters from Mr. Richards and J.A. I assure you I was as well pleased to get them as you could have been after your wait of six long months. The conditions here have continued to improve. We have received more "fresh stuff" since the last boat arrived, and in general the patients are all improving rapidly. Within the past 2 weeks many have started for Valdes. Only 8 remain in my care, and these I hope to get to Valdes within the next 12 days if possible.

If a general relief party can be made up, I will go in charge. I am not so sanguine of success, however. By Saturday next there will be only a half dozen able-bodied men in camp, so assistance will have to be obtained from the Quartz Creek district. It is most difficult to get individuals to make the trip anyhow, although the danger now is not so great as it was in crossing the glacier. Abercrombie's party did not come in on the last boat, but are now expected about the 25th inst. This will be too late to give relief in the way of sending in a party to take the men out as the Klutena River is now opening up, and the trail will soon be unfit to travel.

The body of Dr. Logan, of whom I wrote you, has not yet been found. Five or six of a party of 10 sick men who started from the Chittyna and points below for Orca were frozen on the way. The season has now opened up, so there will be no more such fatalities to record.

The government party is going to cut trail through to the Tanana R., we understand, and several military posts are to be established. Some rather favorable reports from the Teikel and Quartz Creek country have been received, but I consider them as yet only "signs of spring."

A party of men from Cooks Inlet[112] passed above here a short time ago on their way to the headwaters of the Tanana. They have pulled their outfits over, having traveled, counting the relays, about 1000 miles in reaching the Copper River. During this time they camped on the trail, etc., with only a tent for shelter and the weather being perhaps -20° to -50°. Nothing now has been found in the Cooks Inlet region. These men report the death by drowning of a number of men from this side who packed across during the summer.

I have read with interest the prospectus of the Keystone Co., and in general it is good. The statements made by those quoted in it, however, are <u>false</u>, and were given out to deceive people into coming into this season.[113] They have failed utterly. As yet there has been no "pay dirt" found in this region. The country will be more thoroughly prospected this season, and something may be struck. If such is the case, I hope to get in on it; for I have friends who will look after my interests in the Tanana country and about the headwaters of the Copper if it can be reached. If I go out to Valdes I want to return to the Teikel and investigate myself.

A note from Mr. Richards at Valdes tells me he did not get out on the last boat, and he may remain some time there. He hopes to get work with the government party. With regard to our Valdes property, another townsite has been located over ours, and litigation will probably follow if the land proves of value.[114] What do you think of the claim against the government for remuneration for the hospital service? This service is worth at least $2000, but I suppose there is little

---

[112]*Cooks Inlet*: Cook Inlet is on the south coast of the Alaska mainland, N of Valdez. (AP)

[113]*statements ... are <u>false</u> ... this season*: This comment could be in response to a newspaper article that addressed the progress of the Keystone Company. The source of this reference has not been located. (PJT, AP)

[114]*townsite ... litigation*: In Jim and Nancy Lethcoe's *Valdez Gold Rush Trails of 1898-99*, the authors discuss the townsite dispute. (Valdez: Prince William Sound Books, 1996. 28-31) (PR)

chance of any returns. However, the work had to be done for humanity's sake. The service has saved the lives of many.

The news from home is all of interest, though much is surprising. I am made happy, though, in learning that you are all well. I shall shortly write again, and I want to answer all the letters received.

Mother's favors are nearly always read first. Some day I want to step in before her and in a pleasant surprise relieve all her anxiety.

With love to all,
Leroy

Just as the above is finished, a party in from Valdes holds up to my astonished gaze – 3 fresh eggs. Think of it, eggs. I am going to enter into negotiations for one at once.

Leroy

Letter 29 [No envelope]

Valdez,[115] Alaska,
May 2, 1899

By the above you will see that I am in Valdez. We succeeded in making arrangements to bring out the sick from the hospital at Copper Center, as I wrote you we were desirous of doing. It was a hard undertaking, however; and I should not desire to repeat it. In fact there is hardly any consideration that would cause me to do so.

We left Copper Center, Monday, April 17, at 5 a.m. We succeeded in getting the assistance of three Indians and had the help of three men from the Teikel. This, with the hospital corps of four, made the relief party number ten. We were sledding five patients and duffel, with provision, camp equipage, etc.; so you will readily appreciate that our task was no light one.

The details of the trip I shall keep until I am able to relate them in person. Briefly: the first day we made 20 miles, camped at Cox's Landing, about worn out. The next day we made the foot of Klutena Lake, about 12 miles. I stopped at my cabin at the Rapids[116] to pick up a few personal effects, which I desired to take out, and followed the party a little later to the Lake. Assistance in my place from the Rapids to the lake was given by two friends of mine. We stayed at Nokes' restaurant over night and made an early start the next morning across the Lake. Our charges were lessened by one from this point, one of the sick deciding to remain at Nokes. This was the result of a few "words" with me. (The lack of appreciation shown by the sick has in some cases been

[115]*Valdez*: In the beginning of Dr. Townsend's correspondence, the spelling is *Valdes*; when he writes Letter 29 in May of 1899, upon returning from CC, the spelling evolves to *Valdez*. Joe Leahy, Director of the Valdez Museum and Historical Archive, has a tentative explanation: the change in spelling may have been an aide to a shift in pronunciation: *Val· deez*, as pronounced today, as opposed to *Val·dehz*, the common traditional form. The pronunciation in early days, however, is unknown. The change in spelling may have been an outgrowth of anti-Spanish sentiment during the Spanish American War (April 25 to July 17, 1898) and not wanting to be associated with a Spanish name. Joe also believes the "z" may have made more sense than "s" in terms of American pronunciation of a Spanish name (despite how words like "cause" or "does" are pronounced) and may have influenced spelling, which helped foster a shift in pronunciation. The location was named Puerto de Valdés in 1791 (Phillips). (AP)

[116]*the Rapids*: Formerly Amy's Landing. (PR)

more than I could stand. 90% of individuals are absolutely without any sense of honor whatever.) Wednesday we crossed the lake and stopped at a restaurant at the head. Thursday we got an early start and made Sawmill Camp. The trail to this point was bad, and at one place we had to wade the river. We had no gum-boots, so it devolved on 3 of us to plow through just as we were and carry over the others with duffel, etc. Fortunately we were not far from our camping place when this occurred.

The sick were now permanently quartered in a cabin and left with the hospital staff, while I with two Indians continued to the foot of the glacier. These two Indians had been induced to go over the glacier with me; the others went back from the sawmill. My own object was to proceed to Valdes and organize a larger party to bring the men over the glacier, our force being inadequate. At the foot of the glacier we expected to find a restaurant; but finding a couple of deserted tents, I was led to believe everybody had gone. The two Siwash[117] and I therefore put up with very poor accommodations over night, I can tell you. We had but two blankets, and I lay on two washboard pack saddles all night with the result that I was all creased in the morning. Sleep was out of the question. The wind blew a gale, threatening to blow down the tents. I was equally as hungry as when with Mr. Richards we "turned in" about this place, too.[118]

Morning came at last, and you may judge with what disgust I looked a little further up the trail and saw smoke issuing from the restaurant tent. It was dull and heavy the evening before when we came up, and we could not see it. We immediately went up and had breakfast. It was storming on the glacier and impossible to cross. The two following days were also stormy, and we were forced to remain inactive. A Mr. Knott, who had preceded us, was also stormbound here. Monday, the 24th the weather appeared favourable, and we decided to start. Accordingly Mr. Knott, myself, and the two Siwash set out. It was a snow-shoeing proposition all the way and through 6" to 2 feet of snow. Mr. Knott had gone ahead of us; and about noon as we were nearing the summit, we met him returning. He had been to the summit and was inclined to think the weather too bad to continue, a storm having commenced. I urged him to come with us, however, and he consented. We soon reached the summit; and though the outlook was anything but pleasant, determined to push on. The trail blazing proposition became harder, and it was only at intervals that we could get our bearings. I soon became too exhausted to break trail, and the Indians had to take the lead. They were pulling about 125#. Mr. Knott had about 100#. We only reached the 4th bench by 9 p.m. I saw it would be impossible to reach the relief station, so decided to camp in the snow. This 4th bench is where the 7 perished of whom I wrote you. As we pulled through here, we got among the crevasses, having gone too far to the left. Fortune favored us, though; for the cloud lifted, and we were able to get ourselves located.

---

[117]*the two Siwash*: Natives. Originally, jargon for a Chinook Indian, a North American French dialectical form of the French *sauvage* (savage). An Indian, specifically of the North Pacific Coast, a frequent attribution for. Now considered pejorative. (AP)

[118]*I was equally ...this place, too*: The comparison in this sentence is vague: "I was equally as hungry as when Mr. Richards and I stayed overnight at this place once before"? (AP)

Crevasses on the northern side of Valdez Glacier. After straying to the left toward these, fortunately for Townsend, Knott, and the two Natives, the weather cleared some, and their error was seen. (Roppel-DeArmond Col., # 134, circa 1898)

We lay on the snow all night, I with an Indian on either side to keep me warm. We had 3 blankets. I may have gotten a little sleep but doubt it. During the night it snowed. We got off early in the morning and by noon had reached the Relief station. Here we got a cup of tea. I was tired and sore and concluded to remain over night. One Indian, also exhausted, desired to remain; so we made arrangements to put in the night comfortably. Mr. Knott and the other Indian were to continue on to Valdes and return the next day with food and another pair of snow shoes, mine having worn out. They cached their sleds before starting.

Shortly after their departure 3 men from Valdes put in their appearance. They had come for a load of stuff cached a few days before. These men reported the arrival of the government expedition at Valdes a few days before. A little later 3 more men came in about exhausted from the other side of the glacier. They had had a hard trip. These three remained over night. The wind rose to a gale during the night, but it was not very cold. About 2 a.m. I was awakened from a peaceful sleep by a banging on the door and someone calling for Dr. Townsend. I asked what was wanted and was informed that it was a relief party sent for me by Capt. Abercrombie, and they wanted to take me to Valdes. They had provisions of all kinds, of which we all partook. I thus became a case of a relief party relieved. Mr. Knott and the Indian got into Valdez, played out, and reported me at the Relief Station to Capt A., who immediately – 10 o'clock – sent out 2 men and a dog team to my assistance.[119] I was brought on the sled as far as the 3d bench and from there walked to Valdez, where I arrived about 6 a.m.

Captain Abercrombie back in Valdez for the 1899 season. (Abercrombie 1900)

[119]Brown sent out John Fohlin and a man named Johnson with a dog team and provisions. According to Brown they reached the Relief Station about 4:30 in the morning and after breakfast started back, arriving with Townsend and the Indian about 10 a.m. on the 26th. (PR, Appendix A)

"Dog Master [John] Fohlin and his charges at Construction Camp No. 2. June 19, 1899" (caption by Ambercrombie for this photograph in his 1900 Report). Fohlin came for Townsend at the Dempsy Relief station, and he went to the other side of the glaciers after Townsend's patients. (Abercrombie 1900)

Wednesday the 26th. I went to our building, routed out Clark, breakfasted again, etc. Later I called on Capt. Abercrombie and dined with him. I have been royally treated since my arrival here. Valdez is quite a town now. I hope our claim to the town site is being pushed. Those who have located the new site over our claim are working hard for the grant. The government expedition has started their work of trail construction. Even if no gold is found in the interior, this town site will prove of value if a trail is cut through to 40 Mile.[120]

I had hoped to get a place with Captain "A" as surgeon, but unfortunately the very reports sent out by me resulted in his bringing in one, and I am disappointed. Friday a relief party was sent over to bring out the patients at the sawmill. They got in Saturday in good condition with all the sick, whom I have turned over to the government physician here.

I am now relieved of the responsibility that devolved upon me for almost 3 months past and am thankful. I do not yet know whether we may look for any remuneration for the service, but I feel that our claim should receive attention. I am at a loss to know now what is best to do. There

[120]*40 Mile*: Forty-Mile River. (AP)

is nothing open to me that I feel like accepting from the gov. expedition. They are hiring men at $1 per day and board, some $50 a month and board. With nothing prospective, this is N.G. [No Good]. I may go back into the interior, yet the prospects hardly justify. Perhaps I shall soon be on my way home. I do not know. Richards got out on the last boat two days before I got in. I have about $75 in cash now and some owing me, if I can collect. This is enough for the present, so Mother need not worry. I have her letter of March 14th. It was waiting here on my coming. How thankful I was to get it. It is full of interest, and I will answer it so to speak of its contents.

"Valdez, though, in general is dull," writes Townsend. Men who had wintered in Valdez and the Copper River Valley await something to turn up. Captain Abercrombie "gave notice that he would give all idle men employment at the rate of $1 a day for thirty-five days, thus enabling them to pay their steamer fare back to the states. Leaving each with $5 in cash, so they would not be 'amenable to the vagrancy law...'" (SPI, July 7, 1899; see Conclusion). Here "idle" men lounge in front of a barber shop in "old" Valdez, with the harbor in the background, none appearing too eager to cut lumber or work on road gangs for $1 a day. (Snowden Col., Townsend, May or June, 1899)

Dr. Townsend is probably in the company of former patients, two showing the effects of scurvy. The man R of Townsend has black circles around his eyes, and the face of the man before Townsend has a dark cast (see Appendix B for symptoms of scurvy). The countenance of Townsend betrays concern though probably not for his former charges but rather of dismay over his prospects. The four stand in front of the Keystone Company General Store. (Snowden Col., May 1899)

Tonight I dine with Capt. Abercrombie, and perhaps I may learn something of the hospital matter.

May 3d

I spent a very pleasant evening with the Capt., but the hospital question did not come up. I do not wish to push the matter, preferring to bide my time. I hope to have something regarding the matter from you in the next mail.[121] A boat is expected in about the 15th.

The government forces are active just at present, and I believe much will be accomplished this season here. Valdez, though, in general is dull. I wonder how the Col. is getting along. Mother writes of his leaving Seattle.

May 5 – 99

A boat is coming in for mail. Will write again soon.

Love to all,
Leroy

Letter 30 [No envelope—THE LAST LETTER]

Valdez, Alaska.
May 7, 1899

My Dear Mother:–

I last Friday forwarded a letter home telling you of my trip over the glacier and in general of the conditions here at present. I hope you will receive it in due time. As yet I have made no plans; but I do not contemplate returning on the next boat; and perhaps before the arrival of the second, something may present to my advantage. Today I had a talk with Capt. Abercrombie. He expressed himself of the opinion that my claim for government remuneration would receive attention, and offered himself to endorse it. How has Senator Quay succeeded in settling his late difficulties?[122] The Senator is a close friend of the captain, I think, and through him there should be no trouble in getting the claim allowed. The service is worth two or three thousand dollars. If we could get this, what a help it would be. You all might have a little more pleasure then.

Is Elizabeth going out any more? I hope she will. She has her music, dances nicely, and should have a pleasant time. I am pleased at Aunt Lou's getting the carpet; and I hope George will not stop there, but will order the heater in, too. If this claim of mine is allowed, I'll invest part of it in mahogany parlor suites, carpets, etc. that will open your eyes, all of you.

You speak of Father giving up his smoking. I must tell you I am now a lover of the weed myself. You should see my collection of pipes.

The weather here for several days past has been dull and heavy-depressing.

I have wished for a May day at home instead of the everlasting snow. But they [such May days] will come again. Keep up good courage, Mother, and the days will bring your loving

Leroy

---

[121]*something regarding the matter ... next mail*: LST's request for remuneration in consideration of CC hospital service is dated "June 1899" (see Appendix C). This reference could be to a previous request, perhaps to Senator Quay (see next note). (AP)

[122]In 1899 Quay was defeated for re-election by a deadlock existing throughout the legislative session. He was appointed U.S. Senator by the governor to fill the vacancy caused by the failure of the legislature to elect, but the appointment was not recognized by the Senate (Joseph H. Bausnan, *History of Beaver County, Pennsylvania*, 1904, 222). (PR)

Dr. Townsend, an unidentified man, and a puppy at Townsend's feet on the finished porch of the Keystone Company General Store. The front of the Store has been "finished" over the preceding month, but the dismay betrayed by Dr. Townsend's countenance betrays the same concern it did a month earlier. Come the first week in July, he would be back in Seattle, and by July 22, he would be home in Beaver Falls, PA. (Snowden Col., June 1899)

CONCLUSION

by Peggy Jean Townsend

THERE ARE NO OTHER LETTERS in the collection. The thirty letters are tied together; written on the top envelope is the following: "Feb. 3, 1898 – July 13, 1899. 30 Letters (Alaska) L.S.T." The only other item in the packet is a worn piece of letterhead from the Alaska Commercial Hotel in Seattle dated July 13, 1899.[1] So missing is a letter or letters from the period during which Dr. Townsend cared for patients in Copper City. Still, when I acquired the letters in 1969, I did not know about the photographs, to which I gained acess in 1993. Further additions to the collection may surface.

The names of returning prospectors, including that of "L. S. Townsend," made the front page of *The Seattle Times* for Friday Evening, July 7, 1899. The writer gave a telling slant to his story by contrasting the men from Dawson and those from the Copper River region.

> Tales of gold and stories of woe and hardship were brought down by passengers on the steamer *City of Topeka*.
>
> Miners from Dawson staggered down the gang-plank trembling under the weight of the gold sacks on their shoulders. Miners from Copper River staggered also, but not from the weight of gold. They were weakened by a terrible winter in the interior of a desolate country that seems to be entirely barren of the yellow stuff that so many Dawsonites have found.
>
> The Klondikers were able to purchase new suits in the North, and many of them carried dress suit cases. The Copper River men were still in the greasy blanket clothes, and nearly every one had a roll of blankets under his arm.
>
> The contrast was great, and it affected some of the Klondikers. One poor Copper River miner landed here without a cent. His home is in New York and he had to borrow to the fullest extent from all of his friends to get North. He had no hope of getting home. One of the richest Klondikers on board heard of his trouble. On landing he purchased a ticket across the continent, with sleeping car coupon. He fitted the unfortunate out with new clothes and gave him sufficient money to get him through.

The *Seattle Post Intelligencer* for this date also covered the return of men from the Copper River area. It bears reporting for its focus on Townsend and Abercrombie:

---

[1]*a worn piece of letterhead*: The paper is 5 7/8 by 2 3/4 inches, reading "W. F. BUTLER AND SON, PROPS. / COR. FIRST AVENUE AND MAIN STREET. It is likely part of a receipt. (PJT)

COPPER RIVER MINERS RETURN.
Roundly Denounce the Country and Criticize Capt. Abercrombie.

Twenty-one Copper River miners returned on the *City of Topeka* yesterday. There is not a man in the crowd kindly disposed toward the country. With one voice they condemn it as a gold producer. And they unhesitatingly say that in no land under the sun is the prospector met with such hardships. "I consider the Valdes trail the worst trail in any country of which mining history makes mention, and that Valdes glacier is a veritable deathtrap," is the verdict of Dr. Leroy C [S] Townsend, of Beaver [Falls], Pa, and the physician of the Keystone expedition which sought the Copper River clime eighteen months ago.

The scurvy-stricken miners of the Copper River will ever have reason to hold Dr. Townsend in grateful remembrance. He was their ministering angel. Appalling as is the death list, he, the survivors say, saved scores of lives. He went into the country to mine, but devoted his entire time and attention to the sick, receiving nothing in the way of remuneration save the consciousness of having saved the lives of fellow men.

When the scurvy reached its height Dr. Townsend established what is known as the Copper Center hospital, and equipped it with a regular corps of male nurses. To this institution the stricken miners from a radius of many miles about Copper Center were taken for treatment. As fast as they recovered sufficiently to travel Dr. Townsend sent them on to Valdes.

The men who returned on the *Topeka* left Valdes June 25, taking the *Excelsior* for Sitka and Juneau, where they caught the Seattle steamer. They have no words of commendation for Capt. Abercrombie, U. S. A., leader of the government's exploring expedition. Their principle grievance took root in a circular notice extensively posted at Valdes and vicinity. In this the captain gave notice that he would give all idle men employment at the rate of $1 a day for thirty-five days, thus enabling them to pay their steamer fare back to the states. Leaving each with $5 in cash, so they would not be "amenable to the vagrancy law," as the captain is said to have worded his notice.

A. D. Smith, of Washington, is particularly bitter against Capt. Abercrombie, and R. V. Kirkham, a Montana man, says that the captain is doing nothing in the way of rendering real assistance to the Copper River contingent. Smith boldly declares that a sawmill has been established with government funds and that the lumber is being sold for speculative purposes.

Capt. Abercrombie is making poor progress with the trail he is constructing to the interior, according to Kirkham. It is a case of blasting through rocks much of the way, and Kirkham avers that not over a mile of the highway had been completed when the *Excelsior* left. Dr. Townsend, however, expressed the opinion that Capt. Abercrombie would eventually succeed.

About 300 miners remained in the Copper River district when the *Excelsior* left Valdes. There is now no sickness to speak of, the scurvy victims having either recovered or left the country.

Dr. Townsend reports the following deaths, not including the Logan party, as having taken place in the Copper Center section of which he had knowledge:

H. C. Shaw, June 1, 1898
Nick Whan, October [?], 1898
John Dable, December 31, 1898.
[*Andrew Hofler, Dorchester, Rock Co., Wis. Entered Jan. 23d. Died Jan. 28th, 1899.]
S. Milligan, January 31, 1899.
Jack Hayden, January 11, 1899.
Dan Cushman, January 11, 1899.
_____ Mondard, January 11, 1899.
Dan McCullough, February 21, 1899.
Wesley Schnedeger, February, 1899.
A. Lewis, March, 1899.
A. B. Garnes, March 1899. [*A. B. Garnes, Roslyn, Washington. Entered Jan. 23d, 1899. Died Jan. 26th, 1899.]
J. E. Hackett, May [?], 1899.
S. W. Baker, May 9, 1899. [*Samuel W. Baker, San Francisco, Cal. Entered Feb. 5th, 1899, Died March 9th, 1899.]
Andrew Hafter, May 9, 1899.

[*asterisked names appear in an application for remuneration, made to the U.S. War Department, December 19, 1899 (see Appendix C). Of the 23 names listed in this document, 3 died under Dr. Townsend's care, 2 within a few days of being admitted to the Copper Center hospital when Dr. Townsend served there, suggesting the extent of their illness. One died after one month of hospital care. (AP)]

All save Schnedeger, who was frozen to death, died of scurvy. William Willhelm and Alex Stewart are missing, and are believed by their acquaintances to be dead.

The [*Seattle*] *Post-Intelligencer* gave a full account of the death by freezing on Valdes glacier of the members of the Logan party several months ago.

Of the Keystone party of twelve, Dr. Kennedy [Townsend] says all save five have gotten out of the country. One crossed the divide to Dawson, another went to Atlin and three are still in the Copper River section.

≈

How did Leroy Towsend, after having served all that time "without remuneration," or realizing any returns on the Keystone investment, make his way home? The story I have always heard about his getting out of Alaska and back to Beaver Falls was that Lemuel Garrison Townsend, Leroy's father, learned his son had run out of money and was stranded in Valdez. Lemuel asked my grandfather, Charles Walker Townsend, to help with the expense of rescuing Leroy. For that purpose my grandmother, Leanna Wilson Townsend, was persuaded to give up a small inheritance that she had received. C. W. Townsend went to Washington, D.C., and somehow arranged for an expedition to go for Leroy. He was brought back safely. As unlikely as this part of the story seems, it is part of our family lore.

These letters are all that are known to have survived, but the Beaver Falls *Weekly Tribune* during July 1899 tells more of Leroy's Alaska adventures. A reporter compiled the following articles from Leroy's letters and from Seattle newspapers, which had been sent home.

STILL IN SEATTLE

Dr. LeRoy S. Townsend Writes That He Will Remain There for a Few Weeks – The Copper River Not a Gold Producing Country.

A letter received this morning by L. G. Townsend from his son, Dr. LeRoy S. Townsend, dated Seattle, July 6, shows the doctor still there and states that he arrived safely from Valdes by the steamer *City of Topeka*, coming via the inland passage, putting in at Juneau, Wrangle, Port Townsend, etc., the trip being pleasant and one of great scenic beauty. Many Klondikers were on the vessel who had about $200,000 in dust. The doctor does not say when he will start for home as he expects to remain in Seattle for a few weeks. He sent some Seattle papers, from which we copy the following: [quoted above] (Beaver Falls *Weekly Tribune,* July 12, 1899, 5).

S.S. *City of Topeka*. Perhaps in Skagway; the wharf against a steep mountainside suggests Southeast Alaska. This Hegg photograph is taken February 26, 1900, seven months after Townsend's return to Seattle. (Puget Sound Maritime Historical Society, Hegg, # 560-10)

In the July 19 edition, 1899, the following article continues the story:

Sat by a Frozen Corpse

L. G. Townsend on Monday received a letter from his son Dr. LeRoy Townsend, who stated that he would leave Seattle on Saturday last and would stop in St. Paul, Minn., for a few days, expecting to arrive home about Saturday or Sunday of this week. Accompanying the letter was a copy of a Seattle paper in which an account is printed of a death the doctor witnessed while crossing the Valdes glaciers. Charles Johnson and Peter Benson attempted to cross to the coast when the latter broke through the ice and was dragged from the water by his companion. Benson soon froze to death, and Johnson was in almost the same condition when found by Dr. Townsend and others.

The man had gone to sleep near a place where some miners had cached a quantity of provisions, while they returned for other articles. Johnson's life was saved, and he is now carrying mail for the U.S. government. (Beaver Falls *Weekly Tribune,* July 19, 1899, 5)

By the July 26, 1899, issue of the *Weekly Tribune,* Leroy had come home at last, arriving on Saturday, July 22. The newspaper account of his adventures, chiefly in his own words, follows:

RETURNING KLONDIKERS

Dr. Leroy S. Townsend Arrived at Home on Saturday Evening.

NO GOLD IN COPPER RIVER

Dr. Townsend Looks Well and Hearty and His Health Has Evidently Not Been Impaired by the Hardships He Endured – Letters Arrived on Saturday From Seattle Stating That Other Members of the Beaver Mining and Prospecting Company are Returning Home.

Dr. LeRoy Townsend, of this place, arrived at home on Saturday evening from the Copper River district in Alaska, where he went eighteen months ago with Col. Creighton's party in search of gold, but the precious metal was not found in quantities which would pay and the party returned to civilization with the exception of Col. Creighton and three others.

About the time that members of the Creighton Party had decided that gold could not be found in paying quantities in the Copper River district, scurvy broke out among the prospectors at Copper Center and Dr. Townsend put in the winter endeavoring to save the lives of the suffering men. He devoted his time exclusively to the cause of humanity, and how well he succeeded is told in the following brief resume of the trip since leaving home eighteen months ago, given Monday morning by Dr. Townsend.

He said: "We arrived at Valdes on February 26, 1898, and after locating a site for a trading post and erecting a building, we moved our camp equipment and provisions to the foot of Valdes glacier. We commenced the difficult and laborious work of hoisting our stuff with ropes and pulleys to the summit of the glacier, and we had reached the third bench when we were overtaken by a severe wind storm and were almost blown away. The fourth bench put us to the foot of the summit, where we divided into two parties of six men each, one of which crossed the glacier to the timber line on the other side and the second returning for the remainder of our provisions. I was with the first party, which left the summit at 7:20 in the morning, and at midnight we had reached our destination, eight miles away through a heavy snow storm. Our other party was in the worst of the storm and in the snow slide which followed in which a number were killed and from which two bodies were recovered.

"We constructed boats at Seven Mile camp and proceeded onward through Klutena Lake to the head of the rapids, where our party was again divided, and five of our men went on to Copper Center where their tents, provisions, and money were burned.

"We ran additional provisions down to them in boats through rugged canyons over the rapids, a distance of six miles, which was made in fourteen minutes. Mr. Washburn, one of our party from Fitchburg, Mass., weighed three hundred and twenty pounds when we started over the glacier and after taking that ride down the rapids, he had lost in weight enough to reduce his girth measure nine inches, but he expressed a desire to shoot Niagara Falls next.

"It was decided that Hiram Richards and I should go into camp at the head of the rapids on the Klutena and prospect, while the others took the Millard trail to the Mentasta, Forty-two Mile and Sixty Mile Rivers, where they located claims. We made prospecting trips to the Tazlina River and to Quartz creek. On the latter trip I had ten days' rations, and it took me fifteen days to make the trip. I was nearly starved when I came home, and I ate every two hours until I caught up for what I had missed.

"We found no gold in paying quantities, and were ready to go into winter quarters when scurvy broke out at Copper Center, and they sent for me, as one man had died. There were one hundred and twenty-five men sick at one time, and others coming in daily. I had thirty helpless men on my hands at one time and only lost one case. After all but five had recovered I started to take them across the glacier, where we were snow bound for three days. On the fourth day I crossed with two Indians to get assistance. It was a snow shoe trip, with the mountains invisible, the trail obliterated and our shoes sinking 18 inches into the snow. The Indians camped in the snow on the fourth bench, where the storm frightened the Indians so that they trembled like leaves, as they lay on either side of me to keep warm.

"The next day we reached the relief station, where I was compelled to give up, and word was sent to Valdes, and a dog train was sent for me, and relief was sent to the sick men on the glacier. Hiram Richards showed signs of scurvy soon after it became epidemic, and I had him go out to Valdes. He is now at Seattle. One of our men is at Valdes, three are at Forty Mile, where there are good prospects, and Col. Creighton, with two hired men, are in the Atlin country with machinery, where we[2] expect to find some pay." (Beaver Falls *Weekly Tribune,* July 26, 1899, 4)

[2]*we* may refer to members of the Keystone Company, vestiges of the concern still being alive in fact at the time or in the minds of some. (AP)

Surviving correspondence reveals that Leroy did attempt to obtain remuneration for his work with the scurvy patients (see Appendix C). This attempt was not made through Senator Quay, whom Townsend anticipated in his letter of May 7, 1899, would be helpful. Quay was defeated for re-election and thus was out of office.

On December 8, 1899, Abercrombie presented Leroy's claim to the Adjutant General's Office but without success. In the opinion of the War Department, as stated by Judge-Advocate General G. H. Liebor, Leroy's work was undertaken as a private individual, and the provision for the relief of the destitute persons in the Copper River region applied only to relief work done under the supervision of the War Department. Indeed, as Charles Brown, Quartermaster's Agent, shows in his report (Appendix A), "authority" for employment of relief personnel and for purchases is continually cited. The Assistant Secretary of War suggested on January 17, 1900, that Abercrombie determine whether Leroy desired his claim to be referred to Congress. Accordingly, the captain wrote to Leroy, enclosing with his letter all the materials from the War Department.

Meanwhile, on December 19, 1899, Leroy contacted Joseph B. Showalter (Pennsylvania) of the House of Representatives Committee on Labor with his claim. Showalter replied January 26, 1900, stating, "There are over [a] thousand claim bills now in and ahead of the one you might present. There is absolutely no hope of securing passage of a bill in your interest. I am sorry that I can give you no encouragement." (See Appendix C.)

On February 7, 1900, Leroy sent Showalter's letter to Captain Abercrombie in Port Valdez, asking if in light of the Congressman's unfavorable view, Abercrombie still considered it worthy of presentation. Abercrombie on April 3, 1900, wrote, "In view of Honorable Mr. Showalter's letter of January 26th last herewith, I do not now consider it a favorable time to bring this before Congress." Leroy never received compensation for his aid to the sick miners nor remuneration for supplies and medicines acquired for them.

By this time, Leroy had resumed his medical practice in Beaver Falls. The City Directory for 1910 and 1911 reveals that his office was at 1312 Seventh Avenue and that he lived in Patterson Heights. On February 28, 1911, he married Grace Reeves, who had formerly been married to a man named Coates. She had two sons, George W. Coates and Thomas Reeves Coates. Leroy and Grace did not marry until the sons completed their education.

In 1915, the U. S. Navy called Leroy into the service as a lecturer aboard the flag ship U.S.S. *Rhode Island*. One of his hobbies was seamanship, and he became so skilled at navigation (he was reportedly eligible for a master's certificate) that part of his duties while aboard ship was instructing in this subject.

During the hectic days of preparation for World War I, he was commissioned in 1917 as a lieutenant in the Army Medical Corps and worked with X-rays at Magee Hospital in Pittsburgh. After a few months, he was transferred to the Army Medical School in Washington, D.C.

At the close of the war, Leroy was offered three important assignments, finally choosing the post of head of the X-ray department at General Gorgas Hospital in Ancon, Panama Canal Zone. He took up his new duties in 1919, remaining at the hospital for fifteen years. It was shortly after his assignment to Panama that he was given the rank of Major. After fifteen further years of service, Leroy retired from the Army in 1934 and returned to Beaver Falls as a roentgenologist.[1] He became head of the Beaver Valley General Hospital's X-ray department in New Brighton.

[1]*roentgenologist*: Roentgenology, the study and use of X-rays, especially in connection with the diagnosis and treatment of disease. (AP)

During his adventures in Alaska, Leroy showed his interest in guns and hunting, and over the years he amassed a large collection of rifles and revolvers. During his free time he was involved in many firearm organizations: the Balboa Gun Club, of which he was president, the Beaver Valley Gun Club, the U.S. Revolver Association, the Beaver County Buck Tails, the Muzzle Loading Rifle Association, the Amateur Trapshooting Club, and the Mallet Club. His expert marksmanship won him numerous trophies and recognitions.

My family remembers Leroy as very handsome, very big, unusually strong, and in later years, possessing exceedingly white hair. His granddaughter, Ann Coates Wilson, remarked on the pinkness of his face. She remembers him as dressed in a soft beige cowboy hat and matching suit. "He was an outdoor type, jolly and pleasant." His grandson's wife, Jean Coates, remembers his white handlebar mustache and that he liked to wear Panama hats and suits. She thought he looked like Santa Claus.

My father, Charles W. Townsend II, proclaimed Leroy the "most handsome man I ever saw" and one of the strongest. C. W. II told how once on Seventh Avenue in Beaver Falls, when a streetcar conductor insulted a woman who was boarding, Leroy made an eloquent statement to the man without uttering a word: Leroy picked up the conductor from the streetcar, set him on the street, and proceeded down Seventh Avenue without a backward glance.

Dr. Leroy Stewart Townsend died August 24, 1936. He had fallen ill while on the job, probably suffering a stroke. The next day his death was front page news:

> Death last night ended the colorful career of Dr. Leroy Stewart Townsend, 66, retired Army Medical Staff major and distinguished Beaver Falls resident. The eminent X-ray specialist succumbed in the Beaver Valley General hospital, where he had served as roentgenologist....
>
> His profession itself, coupled with his love for travel, had taken him to many parts of the earth's surface – to Alaska during the stirring gold-rush days; to Panama, where he headed an important department in the mammoth Gorgas hospital; and to the South Sea Islands, Australia and South America in numerous interesting tours. (Beaver Falls *News-Tribune*, August 25, 1936, 1, 7)

Not included in the extensive obituary nor in any letters is a story attributed to Leroy about the meanest man he ever met. If true, it certainly signifies the extent to which Leroy's experience in Alaska was interesting. As has come down through the family, Leroy was taking a seriously ill scurvy patient back to Valdez by sled over the glacier. As the man grew continually worse and hope of getting him out alive faded, Leroy, who was exhausted, comforted himself with the thought that he would "inherit" the twenty dollars the man was carrying. However, just before the patient died, he swallowed the money to prevent Leroy or anyone else from getting it.

✱

APPENDIX

APPENDIX A

REPORT OF CHARLES BROWN, QUARTERMASTER'S AGENT.[1]

VALDEZ,[2] ALASKA, *February*— [through June 22], *1899.*[3]

Sir: I have the honor to report the following facts and conditions that have prevailed at this camp and the territory adjacent thereto:

On January 18, 1899, I started provisions, etc., underway to Station 24, with the expectation of landing them there in a few days. Since that time, however, the condition of the weather has been such as to prevent any further movement of these goods. The first cache remains 10 miles from Valdez, and am now awaiting every opportunity to move it up nearer the station. Soft snows, with the river open, followed by severe storms, prevented further progress, though indications are that we can now again safely resume work.

Many parties have of late been crossing the glacier, coming into Valdez from the interior for a change of food, etc. They report the condition of affairs there as being terrible; scurvy being the chief affliction, with many cases of frozen hands and feet. A number have died, and several with frozen limbs will lose them. I have had many requests for aid, medicine, and fresh food from physicians who have patients, and urge immediate action. In addition to those who have come out, it is reported the majority of sick are still in the interior and fear to make the attempt to cross the glacier. Those who have crossed have traveled in large parties for the sake of safety, and have encountered terrific storms.

I have extended such assistance as has been possible, and after much difficulty succeeded in getting some green potatoes and sauerkraut over the glacier. Those who volunteered to take this in found it necessary to make three attempts to cross the glacier before they were able to do so, owing to the trail and storms. Two men died with scurvy while being brought over. They started from Saw Mill Camp and succumbed when near Twelve-Mile Camp. It is reported that 75 cases still exist between the glacier and Copper Center, many of whom will never recover. Those now coming into camp are mostly people who have been lightly touched with scurvy and able to reach Valdez without assistance. Their condition improves after reaching here, and many of them are waiting with the hope of earning something this spring by packing. Some of the number will return to the interior later, while others are anxious to return to their homes though without the means to do so.

On February 23 a party of 25 started over the glacier, after being in Valdez long enough to fully recover their health. On the same day the "Poppy" party arrived here, most of them being frozen slightly about the hands and wrists. After being six days on the glacier they encountered a heavy storm and were compelled to dig a hole in the snow for shelter. On the summer trail Pete Benson lost his life in falling into a creek, being frozen to death before he could be removed to a place of shelter.

---

[1]This report is published in Abercrombie, W. R. *Copper River Exploring Expedition, 1899*. Washington: Government Printing Office, 1900, 37-43.

[2]*VALDEZ* is also be spelled *Valdes* in this 1900 report, written in 1899, the year of transitionary spelling.

[3]*February—[through June 22], 1899*: "June 22" is indicated near the end of this report.

On February 26 W. E. Hendricks and George Hansen arrived here from Copper Center, bringing the report of the death of Robert Izatt, of Milwaukee, Wis; and also the further information that 27 men with scurvy of varying degrees of severity were bound here from the Chestochena[4] and Saling[5] rivers, and that Indians were bringing in two men from the Gakona River.[6]

The Lower Copper River has not been exempt from scurvy. C. R. Waite reports a party en route to Alaganik with six patients afflicted with that disease; four of the number died on the way. Another party following their trail from the Bremmer River had five cases of scurvy with them. February 24 intelligence was received that Dr. Logan was camped at the foot of the glacier with two bad cases of scurvy, the afflicted persons being Adolph Ehrhardt and Max Miller.

March 3 a relief party, consisting of Dempsey and Jackson, arrived here from the glacier with the bodies of [A. E.] Ehrhardt, [Maximillian] Miller, Adolph [August] Schultz, [Rudolph] Eblerkamp [Ellerkamp], and [Alfred] Alleman [Elliman].[7] Ellerkamp's head was crushed[8] when found, and the body of Elliman was found lying across the body of Ellerkamp. Adolph [August] Schultz was a resident of New York, and Ellerkamp formerly lived in Louisville, Ky.

I respectfully request that I may be authorized to rent a house or cabin for the use of destitute prospectors arriving here sick and to whom I could not otherwise give shelter suitable to their condition. I would further ask that the rental date from November 5, 1898, and at a rate not to exceed $10 per month.

I wish to call your attention to the great amount of sickness, particularly scurvy, prevailing in all the camps from Twelve-Mile to Copper Center and through the interior. Many appeals have been made to me from there for assistance, medicines, etc. The following letter, one of a number received of similar purport, may give you some idea of the condition of affairs there:

> COPPER CENTER, ALASKA, *February 3, 1899.*
>
> DEAR SIR: I beg to report to you herewith the serious condition which prevails in this camp. Scurvy has developed to an alarming extent. Two deaths have occurred and the hospital, which it was necessary to establish, is now full, and still throughout the camp are many who should be admitted. Up the Copper River the condition, I understand, is equally serious. Many are wholly without means and dependent upon others for nursing, medical attention, etc. We are in need, too, of some medicines, and of any fresh fruit or vegetables which may be had at Valdez. Two have volunteered their services to bring in these supplies and are now on their way to Valdez. Several men are

---

[4]*Chestochena* River: Chistochina, NE of Glennallen, a tributary of the CR. (AP)

[5]*Saling* River: Likely a typographical error for the *Slana* River; lies E of the Chistochina R, also a tributary of the CR. (AP)

[6]*Gakona River*: proceeds S from the Gakona Glacier to the CR. (AP)

[7]Names in brackets are as they appear in the Coroner's Inquest (see Appendix D). The "Ellerkamp" and "Elliman" spellings are employed hereafter.

[8]"head was crushed": According to the Coroner's Inquest (see Appendix D), the injury was not severe.

suffering with frozen feet, and amputations in a number of cases will, no doubt, be necessary. If it be within your power to give any assistance, I would respectfully ask your immediate consideration.

Thanking you in advance for the attention I know this will receive, I am, yours truly,

LEROY TOWNSEND, M. D.

THE QUARTERMASTER'S, AGENT, *Valdez*.

I have been unable to comply with these requests for medicines, having none on hand, even for use in this camp. If you could ship at once the medicines requested by Dr. Townsend in the enclosed lists, they would be of great value and go far toward relieving distress.

On March 6 the storm was still raging. Dr. Logan's body had not been found up to this time, though search has been made for it by different parties ever since. The steamer *Excelsior* came in and will probably go out to-day, providing they can make a landing for passengers.

Referring to the cache 10 miles up Lowe River,[9] as before remarked, I hoped to move it further up as soon as that terrible storm of fourteen days subsided. Two days after the storm it turned warm, so that I could move neither one way or the other for five days. I at once decided to move the cache back to Valdez, and on April 14, took 1,800 pounds of provisions up to the north of the canyon. After that date the snow melted, so that no more sledding could be done from the lake to the canyon.

In my report of February I made mention of 27 destitute and scurvied people, and from March 5 to April 9, inclusive, 277 additional cases were reported.

Many of the parties who came in from the interior to Valdez were so weakened by scurvy they were obliged to come out to get fresh vegetables, etc., they have in caches in the interior.[10]

On the evening of the 25th of April Mr. Knott and an Indian who had started from Copper Center for Valdez informed me about Dr. Townsend and an Indian whom he had left at the relief station, completely worn out and without food for two days. The Indian who came out with Mr. Knott belonged to Dr. Townsend's party. I at once informed you of their condition and was authorized to send relief to them. At 11 o'clock on the same evening, as soon as the crust on the snow had hardened sufficiently to allow travel, I sent John Fohlin and an extra man named Johnson with a dog team and provisions for their relief. They reached the relief station about 3 o-clock the next morning, and, after having breakfast, started back and arrived in Valdez about 10 a.m., on the 26th, with Dr. Townsend and the Indian.

Upon Dr. Townsend reporting to you that five sick men were left at Barrett's Camp,[11] at the foot of the glacier, on the other side of the range, you authorized me to send at once two dog teams of four dogs each to assist in bringing the sick men over, and on 1 o'clock [sic] Thursday morning, April 27, I sent E. Woods and John Fohlin as drivers, together with seven extra men, John D. Lander, A. Crawford, Joe Weckert, jr., Ed. Anderson, Rob Broughton, Rob Coles, and

---

[9]*Lowe River*: SE of Valdez. (AP)

[10]*they have in caches in the interior*: A misplaced clause or idea seemingly repeated in error. (AP)

[11]*Barrett's Camp*: at the foot of Klutina Glacier. (AP)

T. M. Johnson, who arrived at their destination about 7 o'clock p.m. the same day. On the next day, April 28, they were unable to start back on account of a storm. The glacier at this time of the year is something awful to witness, being composed of glaring ice, horribly huge "benches" almost perpendicular and extending to a great height. It requires the services of four or five men to assist a sick person on a sled down the steep "benches" and over strips of ice so narrow that a single misstep would launch one into crevasses hundreds of feet in depth.

Quartermaster's Agent Charles Brown, an empathic, ethical, and apparently physically fit man well qualified to head up responding to requests for relief during 1898-99. (Abercrombie 1900)

On April 29, the storm having abated its fury, the relief party started back bringing the sick men, Schloegel, Hunt, Liljegren, Launert, and McCracken. They arrived at the Dempsey Relief Station about noon and reached Valdez at 7 p.m. of the same day.

On the morning of April 29, James Nucum came to me and reported that his partner, John Gardner, was at the relief station sick with scurvy and without food, and that he was unable to bring him in without assistance. Shortly after half past 10 on the same day, authorized by you, I sent seven men, Gus Johnson, E. L. Lomnes, John Bornd, Bob Ferguson, Charles Morgan, F. H. Randall, and C. Anderson, to the relief station. They arrived there about noon and started back with the sick man, arriving in Valdez about 6 o'clock in the evening of the same day. Two of the men, Randall and Lomnes, assisted the expedition sent to Barnett's Camp on the way to Valdez from Dempsey Relief Station.

Under your general authorization of October 26, 1898, to care for all destitute persons applying for relief, I secured a house of Tress Glaudman at a monthly rental of $10, beginning November 5, 1898.

You authorized me since your arrival to secure a suitable house to be used as a hospital and another one for a cook house. In compliance with such instructions, I have rented a house for hospital purposes from F. G. Bartlett, the rental beginning from April 22, 1899, at $5 per month; and another house to be used as a cook house from Farmer Brown, the rental beginning April 22, 1899, at $5 per month.

I was informed to-day, May 9, that H. Lennert [Launert], one of Dr. Townsend's party, whom we hauled over the glacier from Barrett's Camp to Valdes [sic], reported himself destitute, but was said to have checks of several hundred dollars. I immediately reported his circumstances to you, who together with myself, investigated the case and found that he had checks on deposit to the amount of $2,000 (checks of $100 each) with the North American Bank, of New York City. He also had a certificate of identification, showing his signature and checks to be genuine. One of these checks was cashed by his permission, less 2 1/2 per cent discount for cashing it, leaving him a balance of $97.50. Out of this amount was deducted $50 for hauling him over the glacier and taking care of him, the money to be divided as follows: Twenty-five for Ed. A. Erickson for taking care of him, $12.50 of which Erickson donated to one Ed. Woods, a very sick and destitute man, who has very little hopes of living, but who wishes to return, if possible, to Seattle, the remaining $25 to be distributed as follows: To F. E. Schloegel, one pair of socks, $0.30; R. P. Ferguson, one pair shoes, one pair drawers, $3.75; F. A. Campbell, one pair shoes, $4; C. G. Nelson, one suit underwear, $2; H. Hertzberg, one undershirt, $1.57 1/2; O. Liljegren, one suit underwear, one pair socks, $2.55; B. Arden, one suit underwear, $2.55; James Ikes, one pair shoes, $3; Gus Johnson, one pair pants, $1.50; showing a total distribution of $50 distributed to the most needed and destitute as per your orders.

I am holding the merchant's receipt for the goods bought and the receipts from those who received the goods.

I also have the honor to report that the total rations issued to the sick and destitute from August 5, 1898, to March 29, 1899, is [$]9,675.

I also furnished to the relief station on the glacier from October 10, 1898, to March 29, 1899, 295 pounds of bacon, 475 pounds of hard-tack, and 20 pounds of coffee.

Lumber for the relief station and coal oil for stoves were furnished by the Steam Whaling Company.

The citizens of Valdez contributed all other necessary provisions and articles, including most of the coffee, extract of beef, stoves, candles, tea, salt, pepper, etc.

I also have the honor to report the following facts concerning the fish in Valdez Bay and streams running therein:

The little candle fish[12] came in this year somewhat earlier than usual, being caught here as early as January. This is accounted for by the fact that Valdez Bay was not frozen over at any time in the season and the fresh-water streams and tributaries were open also nearly all winter, running under the snow. These fish came in immense quantities and inhabited every little fresh-water stream in which they could spawn. We caught thousands here daily with little dip nets and also by hand. The run continues until nearly April.

The herring arrived this year in February, but were not so plentiful as last year. We caught a sufficient quantity, however, for our use. They arrived in immense schools, but not nearer than the Islands and Swanport. These fish, unlike the candle fish, spawn in salt water.

Codfish and halibut follow in the wake of the herring, devouring great numbers of them. We caught plenty of cod, but very few halibut, the latter being a deep-water fish and found only on slopy bottoms. Most of the codfish were caught with the incoming tide and on a very short line. We also speared quite a number, finding them in places in water that did not exceed 1 fathom in depth. The halibut are caught near the sawmill, a distance of 9 miles from Valdez. This fish, the cod, does not inhabit fresh-water streams.

Early in July dog salmon and humpbacks literally filled the streams. We speared them and shoveled them out of the streams. We could have thrown them out by hand, but they have large sharp teeth and are quite vicious. They were so abundant that we selected our victims and speared only the choice ones. In spite of the acquired prejudice against this fish as food, they make a dainty dish.

The silver salmon, the fish canned by all the canneries along the coast, are very plentiful here and are caught with seines in great numbers.

The salmon trout[13] come in October and are sought after with great earnestness and activity. They make a most delicious meal. A pole, with two or three hooks fastened securely on the end, is all that is required to draw them out as fast as is desired. They are the prettiest speckled beauty we have ever seen. The sun shining upon them brings out their brilliant colors and shows them in all their beauty. They sport in their beautiful colors till winter closes the streams.

The first part of May was occupied in cutting and rafting wood across the bay to be used as fuel to the hospital, cook house, and camp in general. Gravel was also hauled and a walk built between the office and the storehouse. During the month repairs were made to the smokehouse,

---

[12]*candle fish*: Candlefish, a small, oily, edible smelt fish; when dried, it can be burned as a candle. (AP)

[13]*salmon trout*: Any of various trouts, such as the steelhead, a rainbow trout that has returned from the sea. (AP)

consisting of the removal of the canvas roof, which was replaced by a shingle roof. The following material was used: Two windows, lumber for the floor, 275 feet; eighteen rafters, 2 by 16 feet, 288 feet; sheeting and finishing lumber, 350 feet; 5,000 shingles; sills and floor joists, 200 feet.

On May 13 we unloaded the Government goods from the steamer *Excelsior*.

We took the roof off the stable and raised it 2 feet higher, replacing it with a shingle roof. The amount of material used was 23,000 shingles; 60 rafters, 2 by 16, 16 and 18 feet long; 3,000 feet of sheeting; gable ends, 400 feet; lumber for floor, 2 by 10, 4,000 feet; mud sills, 6 by 6, 1,280 feet. The floor upstairs remains unfinished.

We built station No. 2, 15 by 30 feet; hauled logs and lumber for the station. The amount of material used was 65 logs; 450 feet of lumber for flooring; 450 feet of lumber for gable ends and finishing; 9,000 shingles; 36 rafters, 2 by 6, 14 feet long; 800 feet sheeting. Hewed logs for floor joists. The station was built 2 feet above the ground. A wagon road was cut from station No. 2 to the foothills where the road takes the bar to Keystone Canyon. A trail was cut for surveying purposes. The Government steam launch was furnished with water and fuel whenever necessary.

On June 13 we removed the hospital to Swanport; took the lighter to the sawmill for a load of lumber and slabs for Valdez; cut trail from the end of wagon road for pack team, to avoid a deep stream.

June 22 we went to Swanport to cut firewood and piles and sawlogs and continued this work to present date. Fifty logs were rafted to the sawmill. Two hundred and one piles, from 30 to 50 feet long, were cut, peeled, and made ready for use. We also have on hand fifty logs for firewood.

The mowing machine and rake were taken from Valdez to the island to be used in putting up hay.

With reference to affairs at the hospital, Oscar Liljigren [Liljegren] is going out on the first boat. John Gardner and H. E. Hunt are the only ones remaining. Hunt, with his frozen foot, is still black with scurvy from the ankle down. It will be necessary to amputate his big toes, but, on account of scurvy, the operation must be delayed.

Schloegel and Gardner still have scurvy from the ankle down, but are on the way to recovery. They refuse to be put on the pay roll on account of not being able to do anything in the [l]east and have plenty of supplies in the interior. Schloegel's supplies are on the Christochina River, about 200 miles distant from Port Valdez. Gardner's supplies are at Copper Center.

Respectfully submitted.

CHARLES BROWN,
Quartermaster's Agent.

Capt. W. R. ABERCROMBIE,
Second United States Infantry,
Cmdg. Alaska Exploration and Relief Expedition No. 2.

APPENDIX B

REPORT OF LEROY J.[1] TOWNSEND, M. D., ON SCORBUTUS, OR SCURVY.[2]

Scorbutus, or Scurvy, the most dreaded disease of the Alaskan prospector or miner, is a constitutional disorder, characterized by a vitiated state of the blood dependent upon the lack of necessary elements supplied through fresh fruit and vegetable nutriment. This lack or insufficiency in fresh vegetable material is the essential factor in its development. Another factor is the long-continued use of salt and smoked meats; the use of stale or unwholesome food still another. Indolence and inactivity are predisposing, and mental lethargy or depression is undoubtedly active in its production.

The changes in the composition of the blood show a marked diminution of the potash salts, and, contrary to a quite general belief, the blood is not thick, but thin and watery. Indeed, to this watery condition may be attributed the discolorations which manifest themselves during the disease, the liquid condition of the blood admitting of its passage through the vessel walls into and beneath the skin. The development of the disease is usually slow and insidious. The unfortunate suffers from malaise and is indisposed to either mental or physical activity. Shortness of breath follows the slightest exertion, and palpitation of the heart is of frequent occurrence. Vertigo may be associated. Pain occurs in the knees or in the muscles of the calf or thigh, the muscles so contracting as to produce lameness. Œdema[3] of the extremities, particularly marked about the ankles and along the course of the tibia or shin bone, is an almost constant symptom. Puffiness of the face may be apparent. The pitting of the parts on pressure is marked. The skin is dry and harsh and presents extravasations[4] of blood, in size from petechial[5] spots to almost the whole surface of the part. These discolorations vary in color from a light red to a dark blue or black. Desquamation[6] may follow over the ecchymotic[7] surfaces. There are swelling and discoloration of the gums, which tend to bleed readily. The swelling may greatly interfere with the ingestion of food. The teeth loosen or drop out. The breath is terribly offensive. General weakness and depression are progressive, with decided emaciation. Anæmia is marked; the face pale, yellowish, or leaden in color. The eyes are shrunken and surrounded by dark circles, and in cases of long duration a characteristic odor has seemed to me to exist. The urine is high colored, abnormal in quantity in many cases, and loaded, becoming highly offensive if left standing. Hemorrhage may take place from various mucous surfaces. Fever of a remittent character[8] is a

---

[1]The "J." is in error; it should be "S," for Stewart. (PJT)

[2]This report is published in Abercrombie, W. R. *Copper River Exploring Expedition, 1899*. Washington: Government Printing Office, 1900, 44-47.

[3]*Œdema*: Edema, abnormal retention of fluids. (AP)

[4]*extravasations*: Extravasation, the escape of an organic fluid (e.g., blood, sap) from its proper vessels into the surrounding tissues. (AP)

[5]*petechial*: Of the nature of petechia: a small red or purple spot in the skin caused by extravasation of blood, occurring in certain fevers. (AP)

[6]*Desquamation*: The coming off in scales or scaly patches, especially that of the epidermis as a result of certain diseases; peeling. (AP)

[7]*ecchymotic*: of the nature or appearance of an eccymosis: a blotch caused by extravasation of blood below the skin. (AP)

[8]*Fever of a remittent character*: Remitting, or abating, for awhile or at intervals and then returning. (AP)

constant symptom, the morning temperature usually registering 99 1/2° to 100 1/2°. In the afternoon it reaches a degree or two higher. A good appetite usually exists, at least until the disease is far developed.

In such cases as progress unfavorably, the weakness increases, the patient becoming unable to assist himself in the slightest degree. He soon lapses into a semicomatose or comatose condition, from which death relieves him.

Such complications as I have seen have involved the lungs and kidneys. I have had no case in which scorbutic dysentery has followed.

In diagnosis it would seem impossible to confound the disease with any other affection, especially when the condition is well developed. Yet the most astonishing errors were made in the Copper River region during the past year. Nor were these errors made by the laymen alone, but by presumably reputable physicians. For instance, one case that came to my notice had been diagnosed and treated as gangrene of the feet and legs, thought to have been produced by too tightly constricting the parts with cords used in tying gunny sacks over the feet and limbs. The marked extravasations were responsible for this mistake. The patient died.[9] Again, another individual who had been unfortunate enough to have his toes frozen had the same cause placed on these offending members in explanation of scorbutic symptoms, which subsequently developed. Another individual, in which the disease was just beginning to manifest itself, was told that the pain and stiffness in his leg were due to a sprain, and the slightest swelling and discoloration of the gums were the result of "frosting" them.

In general, through the opinion expressed by several physicians, the impression was that the condition was rheumatic. Blackleg rheumatism it was called. This was a most unfortunate mistake. The line of treatment instituted in these cases was directly antagonistic to the accepted lines followed in scurvy, and such as subsequently came under my care responded very tardily to antiscorbutic measures. Under this impression, too, many had taken the trail who, had they known that the developing pains were the forerunner of scurvy, would not have done so. Many of these unfortunates had to be hauled back, and much suffering and inconvenience were the result. The only disease which might be mistaken is a rare one, known as peliosis rheumatica, or Schonlein's disease.

The above should be sufficient to place anyone on his guard, and, with reasonable discernment, he should be able to recognize the development of this dread malady.

In the consideration of treatment preventive measures should first receive attention. Citric acid lemonade should be used frequently—from one to three times daily. It may be either sweetened or unsweetened, and in strength corresponding with lemonade made from the fresh fruit. Lime juice may be used in the same way. Cider vinegar is of service. Dried or evaporated fruit and vegetable products will not serve as satisfactory substitutes. Canned fruits are of greater value. I would urge, too, that the Alaskan prospector and miner take a plentiful supply of milk, butter, sugar, and eggs (Lamont's crystallized). These articles should not be looked upon as luxuries; they are necessities.

---

[9] *The patient died*: Mr. Grund, Letter 19, May 26, 1898. (AP)

On [sic] the interior a popular preventive and remedy was an infusion of pine needles. This was prepared by selecting limbs of young growth of the pine, stripping off the needles, putting in boiling water and setting aside to steep. Some boiled the needles for a considerable time, making a stronger decoction.[10] Of this tea a fourth to a cupful was taken two or three times daily, as required. The inner bark of the willow and juniper berries were used in the same way. All of the above have proven of value. I did not prescribe them in my practice, but I am satisfied of their usefulness and they are worthy of remembrance.

In direct treatment the indication is to supply as far as possible the principles needed for the healthy constitution of the vital fluid, the blood, and to directly change the environment which tends to the development of the disease. Place the patient in the best possible condition for the invigoration of the system. If possible, let him take gentle exercise. Supply fresh fruit and vegetables, lemons, oranges, apples, potatoes, onions, cabbage, etc., together with fresh meat, if available. Lemons probably give the quickest results, and it is astonishing the improvement that will follow their use. Raw potatoes are excellent. Divert the mind of the patient and inspire hope and cheerfulness.

Tonics are indicated. The tincture of the chloride of iron, the citrate of iron, the citrate of iron and quinine, or the citrate of iron, quinine, and strychnine have proven of the greatest value in my hands. The mineral acids may be used to advantage. For the oral symptoms a wash of potassium chlorate will answer admirably, or one of boracic acid or equal parts of boracic acid and borax, with a few drops of carbolic acid added. Attend to the secretions, keep the bowels open with mild laxative medicines. Relieve pain with anodynes, resorting if necessary to morphine. The use of hot-water bags for relief of pain in the limbs or back is excellent. The application of plasters may be beneficial, as well as hot fermentations. Other symptoms must be met as they arise.

I believe that operative procedure is contraindicated when scorbutus exists. In two cases operated upon the greatest difficulty was experienced in controlling hemorrhage. In neither case was there manifest symptoms at the time of operation, but symptoms developed very soon after. Considering the blood changes, difficulty in stopping hemorrhage might be expected.

To my knowledge no cases of acute rheumatism, of typhoid, cerebrospinal, or other fever developed in the Copper River region.

Considering the hardship and exposure undergone, it is surprising that so little sickness resulted.

LEROY J. [S.] TOWNSEND, M. D.

Valdez, Alaska, *May, 1899.*

---

[10]*decoction*: Action of decocting, especially boiling in water or other liquid to extract soluble parts. (AP)

APPENDIX C

Typescript Record of Leroy S. Townsend Request for Remuneration

Original request, apparently prepared and forwarded at the urging and assistance of Captain Abercrombie, Commanding Officer of the Second United States Infantry, Alaska Explorations and Relief Expedition No. 2:

Port Valdez, Alaska, June, 1899.

United States Government.

To LEROY S. TOWNSEND, M. D., Dr.

To Professional Services:

Medical and Surgical, Medicines, Supplies, etc., as surgeon in charge Copper Center relief hospital, Copper Center, Alaska.
From January 31st, 1899, to April 30th, 1899. $2100

Home Address:
Beaver Falls, Penna.

[page 1 as attachment to the above]

The large number of sick in the Copper River country without means and without anyone to attend them made necessary some measure for systematic relief. Accordingly the Copper Center Hospital was organized January 23d, 1899, with Mr. C. F. Collins, of Portland, Oregon, superintendent, and L. W. Guiteau, of Freeport, Ill., and E. B. Wheat of Ellensburg, Washington, in charge of kitchen.

My service as physician began January 31st and continued until April 30th, 1899. During this term 14 patients were cared for in the hospital, directly, and 7 received attention in an out-door service. The average attendance on the 14 was for 61 days, and on the 7 for 30 days.

Following is a list of patients, copied from the hospital register:

A. B. Garnes, Roslyn, Washington.
Entered Jan. 23d, 1899. Died Jan. 26th, 1899.

Andrew Hofler, Corchester, Rock Co., Wis.
Entered Jan. 23d. Died Jan. 28th, 1899.

Oscar Liljegren, Chicago, Ill.
Entered Jan. 23d. Discharged April 30th, 1899.

H. E. Hunt, Toledo, Washington.
Entered Jan. 24th. Discharged March 25th, 1899.

Charles Gorich Morris, Grundy Co., Ill.
Entered Jan. 25th. Discharged March 25th, 1899.

J. G. McCracken, 63-56 Eggleston Ave., Chicago, Ill.
Entered Jan 25th. Discharged April 12th, 1899.

John Gardiner, Gunnison Co., Colorado.
Entered Jan. 26th. Discharged April 17th, 1899.

[page 2 of attachment]

D. M. Dennie, Madrid, Boone Co., Iowa.
Entered Jan 29th. Discharged April 18th, 1899.

Hiram Cramer, Osceola Mills, Clearfield Co., Pa.
Entered Jan. 30th, 1899. Discharged March 17th, 1899.

C. E. McCoy, 747 61st Place, Sta. O, Chicago, Ill.
Entered Jan. 31st, 1899. Discharged April 8th, 1899.

Samuel W. Baker, San Francisco, Cal.
Entered Feb. 5th, 1899. Died March 9th, 1899.

Edward Young, Port Huron, Mich.
Entered Feb. 6th, 1899. Discharged April 14th, 1899.

Jack C. Paulson, Laramore, North Dakota.
Entered Feb. 8th, 1899. Discharged March 27th, 1899.

Frederick E. A. Schloegel, 2015 E Baltimore St., Baltimore, Md.
Entered Feb. 16th, 1899. Discharged April 30th, 1899.

William N. Lewis, Bellefourche, South Dakota.
Entered Feb. 21st, 1899. Discharged April 3d, 1899.

Philip Straub, 429 S Chestnut St., Los Angeles, Cal.
Entered March 3d, 1899. Discharged April 3d, 1899.

Out-door Service.

Capt. F. M. Marlatt, Lawrence, Kansas.
Entered Feb. 1st, 1899. Discharged April 17th, 1899.

John Thrift, Rock Springs, Wyoming.
Entered Feb. 1st, 1899. Discharged April 3d, 1899.

James L. Hamilton, Portland, Oregon.
Entered Feb. 25th, 1899. Discharged April 3d, 1899.

P. Gillispie, Bakersfield, Cal.
Entered Feb. 25th, 1899. Discharged April 3d, 1899.

Lars Bernard Erickson, Portland, Maine.
Entered March 18th, 1899. Discharged April 3d, 1899.

Nels Peter Jansen, Summit, South Dakota.
Entered March 1st, 1899. Discharged March 28th, 1899.

Alfred Johansen, Biloxi, Miss.
Entered March 21st, 1899. Discharged March 23d, 1899.

[Follow-up inquiry to request for remuneration]

LEROY S. TOWNSEND, M. D.,
BEAVER FALLS, PA.

[Stamped on letter: "A. G. O. {Adjutant General's Office} Div. Received DEC 26, 1899]

Beaver Falls, Pa., Dec. 19th, 1899.

Hon. J. B. Showalter [Representative from Pennsylvania, Committee on Labor]

Washington, D. C.

Dear Sir:--I am just in receipt of a letter from Captain W. R. Abercrombie, 2d Inf., U. S. A., informing me that a claim I am making for professional services rendered sick and destitute prospectors in the Copper River region, Alaska, has by him been forwarded to General Corbin, Adj't Gen'l. U. S. A., War Dep't. Might I ask that you confer with General Corbin relative to the status of the claim, and advise me accordingly as to possibility of allowance or as to further procedure in the matter.

The work for which remuneration is asked comprehends three months service, medical and surgical, with supplies, etc., in a hospital at Copper Center, Alaska, organized for the care of destitute sick in that region during January, February, March, and April of the present year. My term of service began January 31st, and continued until April 30th. During this period 14 patients were cared for directly with an average attendance of 60 days, and 7 patients in our out-door department for an average of 30 days.

This work was undertaken voluntarily, and without any expectation of remuneration, but learning that there had been an appropriation of $30,000 made for the relief of Copper River prospectors, I presented a bill in the sum of $2100 in the hope it might receive the consideration which the service merits.

Captain Abercrombie in charge of the Copper River Government Expedition, assured me of his approval.

Thanking you in advance for the attention I know will be accorded this, and assuring you of my appreciation of any effort in my behalf, I am,

Yours very truly,

Leroy S. Townsend, M.D.

[Upon receipt, in large cursive, Representative Showalter wrote the following:]

Dear General [presumably Corbin]:–
Kindly give me status this claim and oblige.
Sincerely Yours

JBShowalter

Dec 25 1899

[The following, in approximated typescript of cursive, is a reply from Representative Showalter; it appears on a half sheet of 8-X-10-inch paper:]

FIFTY-SIXTH CONGRESS.

JOHN J. GARDENER, N.J., Chairman
James T. McCleary, Minn.
William Lorimer, Ill.
Joseph B. Showalter, Pa.
Richard Bartholdt, Mo.
K. Henry, Conn.
William H. Graham, Pa..
Louis W. Emerson, N.Y.
W. Jasper Talbert. S.C.
John S. Rhea, Ky.
Amos J. Cummings, N.Y.
Ben F. Caldwell, Ill.
Daniel J. Riordon, N.Y.
John G. Shreve, Clerk.

Committee on Labor,

House of Representatives U. S.,

Washington, D. C., Jan 26, 1900

Leroy S. Townsend MD
Dear Doctor:-
There are over thousand Claim bills now in and ahead of one you might present– There is absolutely no hope of securing passage of a bill in your interest– I am sorry that I can give you no encouragement.
Yours Truly JBShowalter

APPENDIX D

Typescript of Coroner's Inquest On Deaths in Dr. Logan's Party

Stenographic report of evidence and proceedings held at Coroner's inquest over the bodies of Rudolph Ellerkamp, Adolf Ehrhardt, August Schultz, Maximillian Miller, and Alfred Elliman, Port Valdez, Alaska, March 4, 1899.

George Haber sworn, said he lived at 134 W. 25th St., New York. I identify Maximillian Miller and Alfred Elliman as two of my partners in the Scientific Mining Co., Elliman is unmarried. I also identify Adolph Ehrhardt who is another of my partners in the Co. Miller and Ehrhardt were both sick with the scurvy. They engaged Dr. Logan, Schultz and Elliman to take them over the Valdez glacier on sleds for the sum of $150. I positively identify these three men as we all camped together at 12-Mile camp. We left New York City all together. Schultz and Ellerkamp I know and identify also as two men camping near ourselves and as the men engaged to bring my late partners over the glacier. Addresses.

Ehrhardt 35 E. 1st St., N.Y.
Elliman, 112 W. 12 St., N.Y. % Miss Mary Bieri.
Miller, 310 E. 80th St., N.Y. % John Wegner.

Otto Kanitz sworn, identified the body of August Schultz, the Vice-President of the Tanana Mining Co. Said Schultz lived in New York. Was well acquainted with the deceased as I traveled and camped for some time with the Tanana Mining Co.

George Leturner[1] sworn, identified Rudolph Ellerkamp. He stated he did not know Ellerkamp's address, but could give it later. He is a correspondent of the Louisville Post. I know nothing of his death, except that I heard the party was coming over the glacier and was caught in a storm. I was one of the relief party sent out. Was present when the two sleds carrying the two sick men were found. Miller and Ehrhardt were on the sleds rolled up in their blankets. Schultz was lying beside them covered up with snow. These three I saw before they were taken out. The other two, Ellerkamp and Elliman were found later by others in a crevasse about a mile this side of the sleds. Jackson, Simpson, Hogg and Williams of the relief party were with me at the time. Dempsey and Anderson were left with the bodies while the rest attempted to find the rest of the missing men.

Mr. Dempsey sworn, stated he was partly instrumental in organizing the relief parties. The first party consisted of five men who brought the bodies down on sleds to the relief station the first night. Williams and I went up later and took out Schultz but could find no more there. Jackson and Leturner started down to the left and I to the right. Two more took the sled with Schultz and walked down the trail, all agreeing to signal if any one was found. I saw some tracks up over the crevasse which I followed up and saw two bodies lying down in a crevasse. We succeeded in getting them out and put them on sleds. One had his face partly covered with blood as though he

[1] *Leturner*: Listed elsewhere as "Latturner." (AP)

had injured himself when falling in. The crevasse was 12 feet deep and probably 15 feet long. These two were found Friday morning about 10 o'clock. They were immediately brought down here. I did not know the deceased personally. We found a few things lying around loose, a list of which Mr. Williams has.

Peter Jackson sworn, said he was one of the relief party and was present when the bodies were recovered. I concur in all that Mr. Dempsey and Leturner have said. I am acquainted with Schultz and Ellerkamp and identify them.

W. J. Nolan sworn, identified the bodies of Ehrhardt, Miller, Ellerkamp, Schultz and Elliman. They were all camped at 12-Mile since 20th of May. Schultz and the still unfound Logan camped near my tent, the others were about 1/4 mile off. They left 12-Mile camp Friday morning and were frozen between Monday and Tuesday night. Was acquainted with Schultz personally.

J. B. Ridge identified Schultz and Ellerkamp. I saw the party on the glacier at a distance. The storm was raging so I could not render them any assistance.

George Hogg sworn, identified Schultz and the two sick men who were found together. I corroborate Mr. Dempsey's evidence in all particulars as I was one of the relief party present.

Charles Anderson also reiterated what Mr. Dempsey had said. I was one of the relief party. Am acquainted by sight but not personally with the deceased.

W. H. Simpson sworn, identified August Schultz. I was one of the relief party. We found Schultz in a snow hole and the two sick men on sleds at the foot of the fourth bench. The other two bodies were in a crevasse about one mile this side. Ellerkamp and Elliman fell off a precipice of ice and remained side by side.

Isaac Williams said he was one of the relief party and assisted in recovering the bodies as described by Mr. Dempsey in his testimony.

Axel B. Elder said he could positively identify Schultz and Ellerkamp being thoroughly acquainted with them. I left the other side of the glacier one hour after the deceased party left, viz. -9 a.m. I overtook them about one mile up the first bench. I went right on along with my sled and reached the relief station that evening.

J. R. Eagins swore that the trustees met in his office in Valdez and appointed a committee to take charge of the bodies and administer on their estates and take charge of their effects. Eagins, Dunham, Smith and Williams were appointed, Dempsey afterward taken in. I afterwards drew out of the committee. The committee now consists of Smith, Williams and Dempsey.

Examination of bodies by Dempsey, Smith and Williams in company with Dr. A. Von Gunther.

Body No. 1--Identified in evidence as that of A. E. Ehrhardt. 35 E. First St., N. Y. All underclothing marked A. E. sewed in. Pad on sweater, two undershirts, overshirt, vest, pants, two pair drawers, socks and moccasins. Found on body fine comb, letter addressed to his wife, 1 box cartridges. Doctor said died from freezing.

Body No. 2.--Maximillian Miller. Nothing on body but a red handkerchief, clothed same as No. 1. No. 1 and two were on sleds when found and presented evidences of having been pulled by the others.

Body No. 3--Identified as August Schultz, president of the Tanana Mining Co. Had plain gold ring on finger. Letter from New York having order of O. of U. F. printed heading. Ring in pocket with initial G. 23 cents in small bag. Bone match box, whistle, pad-lock, watch and chain, $95 in bills and $45 in gold. Money was found on him up on the glacier. Watch has No. 2165263 Waltham works. Doctor declared he had died from freezing.

Body No. 4--Identified as Alfred Elliman. Clothing consisted of sweater, two pair pants, 2 pair drawers, 2 undershirts, 2 overshirts, pair canvas mucklucks. Money belt found next his shirt containing silver $12.50, gold $25.00, paper $137. Doctor declared death caused by freezing. In small sack found on glacier $5.69 silver, $5 gold.

Body No. 5--Identified as that of Rudolph Ellerkamp, correspondent of the Louisville Post, identified by all present who were acquainted. Blood on face from fracture caused by falling in ice crevasse. Found on body, cartridge belt and revolver, suspensory bandage, blue handkerchief, gold watch No. 410766, Keystone watch Co. Belt containing $30 bills, some paper clippings. Had on canvas coat, sweater, blue undershirt, mackinaw pants, canvas mucklucks. Doctor declared death from freezing.

Dr. A. Von Gunther in his statement said: I examined the bodies that were frozen on the glacier, and find as follows: The two bodies on the sleds had scurvy but died from freezing. The other three were in apparent good health and died from freezing also. One of these three identified as Ellerkamp met with an accident as shown by a cut on the nose, probably caused by falling in the ice and then freezing to death. The accident was not serious and nothing to cause death of itself.

Jury Finding.

We find that the deceased bodies of Rudolph Ellerkamp, Maximillian Miller, Alfred Elliman, August Schultz and Adolphe [sic] Ehrhardt met their death by freezing on the glacier about 14 miles from Valdez on 27th of February, 1899.

Inquest was here adjourned.

A. B. Elder, Stenographer.

Appendix E—Documents

The following typescript is of the agreement made between Leroy Townsend and his father and uncle on January 28, 1898, prior to Leroy Townsend's leaving Beaver Falls, PA, for Seattle.

THIS AGREEMENT, made and entered into the Twenty Eighth day of January in the year one thousand eight hundred and ninety eight, by and between L.G. & C. W. Townsend of Beaver Falls State of Pennsylvania, party of the first part, and Leroy S. Townsend of Beaver Falls State of Pennsylvania, party of the second part.

WITNESSETH: That whereas the said parties of the first part have this day paid to the said party of the second part the sum of Two Hundred Fifty ($250⁰⁰) dollars, receipt of which is hereby acknowledged. Therefore this party of the second part proposes and agrees to use said money in outfitting himself to go with the Keystone Party, of said state, to the Copper River Country, or other parts of Alaska, there to locate a mining claim or claims, and to transfer an undivided Five twelfth (5/12) of a one half interest in such claim or claims, to the parties of the first part, and to record same and to deliver proper record to the parties of the first part.

To mine said claim or claims to the best of his ability during the mining season, and to deliver to the said parties of the first part, Five twelfth (5/12) of one half of the net proceeds so mined, for the term of one year from date hereof.

All mining claims located by the Keystone Party are to be consolidated into one property, and each claimholder or claimholders are to receive their proportionate share of the output, and to own his proportionate share of all property acquired.

IN WITNESS WHEREOF, we, the said parties to these presents have hereunto set our hands and seals, the day and year above written.

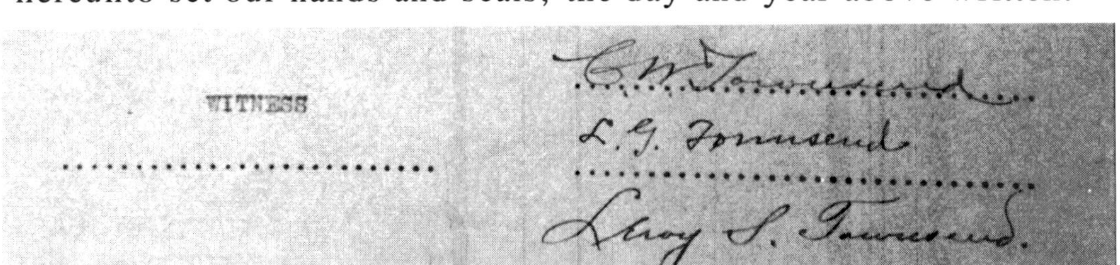
WITNESS

C. W. Townsend

L. G. Townsend

Leroy S. Townsend

E. B. SIMPSON, Secretary. JOHN WARREN, President.

The Keystone Company.

MINING

Rooms 74 and 75 Schmidt B'ld'g, 339 Fifth Avenue.

Pittsburg, Pa. January 5th, '99. 1899.

L. G. Townsend, C. W. Townsend, A. J. Leigh, W./ P. Barnum, and Dr. Wm. S. Cook.

Gentlemen:–

Enclosed you will please find the certificates of stock of this Company to which you are entitled as the representatives of a one-half original interest and for the $75.00 paid in under call of November 21st. viz.

L. G. and C. W. Townsend, 5/12 1000 and 125 shares (125)

A. J. Leigh, 2/12, 400 and 50 shares.

W. P. Barnum, 4/12, 800 and 100 shares.

William S. Cook, 1/12, 2009 and 25 shares.

The other 2400 shares represented by this interest will be issued to Leroy S. Townsend.

We also send you a number of copies of our Prospectus and hope that each of you will make an earnest effort to assist in selling the stock now placed upon the market, for which work by authority of the Board of Directors you will be allowed five per cent on the cash received as commission for selling the same. Please send name and address, giving street number if any plainly written to the secretary and the certificate of stock will be sent to you through Mr. Warren. It will be necessary to make an energetic effort to dispose of this stock

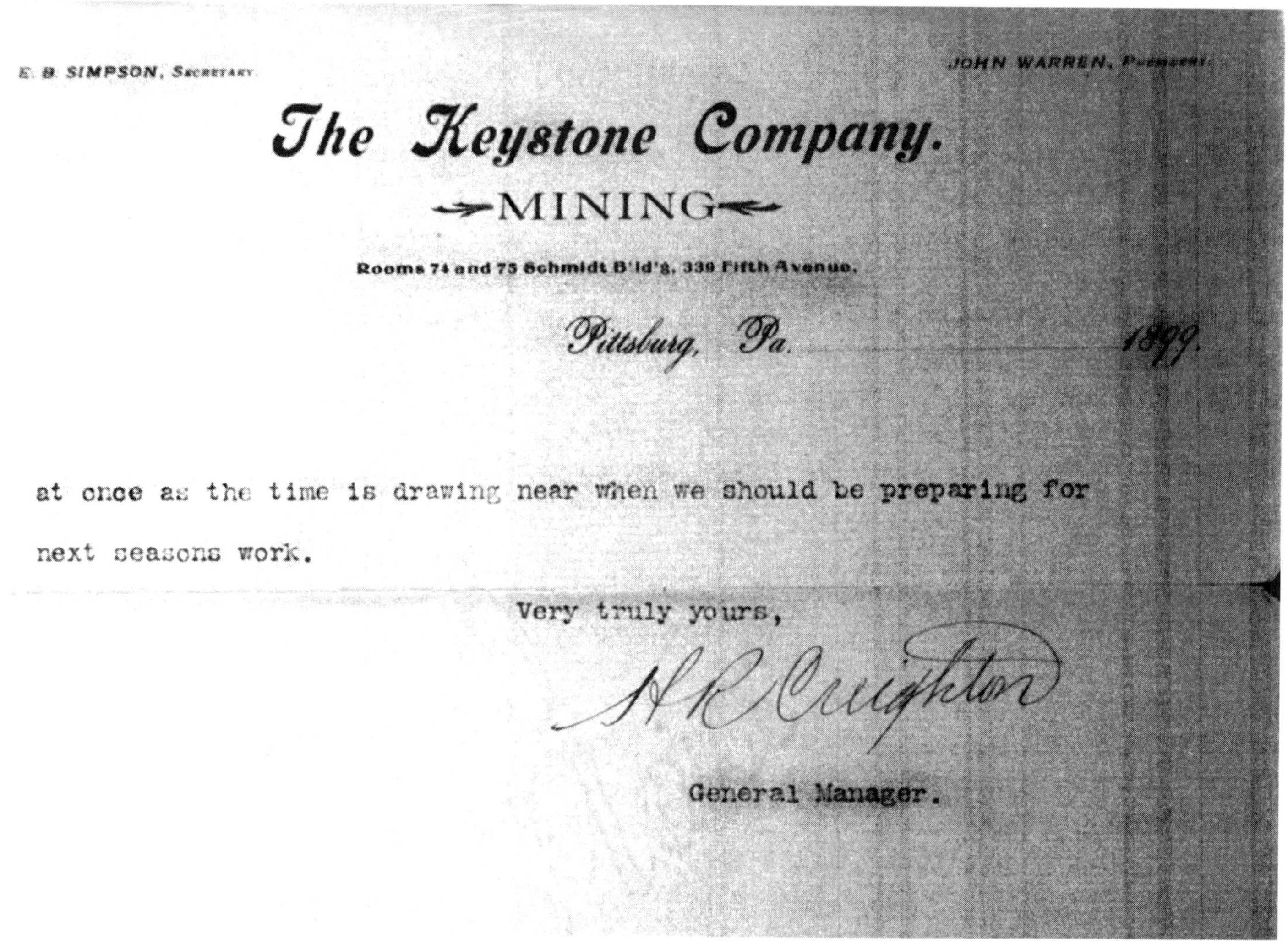

E. B. SIMPSON, SECRETARY. JOHN WARREN, PRESIDENT.

The Keystone Company.

MINING

Rooms 74 and 75 Schmidt B'ld'g, 339 Fifth Avenue,

Pittsburg, Pa. 1899.

at once as the time is drawing near when we should be preparing for next seasons work.

Very truly yours,

H R Creighton

General Manager.

The above document, signed by H. R. Creighton, shows that development of the Company continued. There is no evidence, however, that this stock was sold or that a profit was made (except by way of experience) or even the return of the initial investment.

On the following page appears an agreement between W. J. Woliben and Leroy Townsend over a "Teikel District" claim. Woliben, a Copper Center Hospital volunteer, is given half interest in Townsend's claim in consideration for development of the property. There is no known evidence from the Townsend side of the family that anything came of the arrangement. The document seems hurriedly crafted, and it is likely a second document for Woliben was "fair copied" from the first, for his use. The assignment indicates that Townsend had decided to leave Alaska as of, if not before, June 8, 1899.

Port Valdez, Alaska.
June 8, 1899.

This agreement, made and entered into this eighth day of June, 1899, by and between W. J. Wilobur of Lincoln, Neb., party of the first part, and Leroy J. Townsend, M.D., of Beaver Falls, Pa., party of the second part, Witnesseth: That the party of the first hereby agrees to do all necessary work to secure the right and title of the party of the second part to a certain property, known as the Mulligan Claim, and situated in the Tiekel District, Alaska, subject to agreement entered into by original locators.

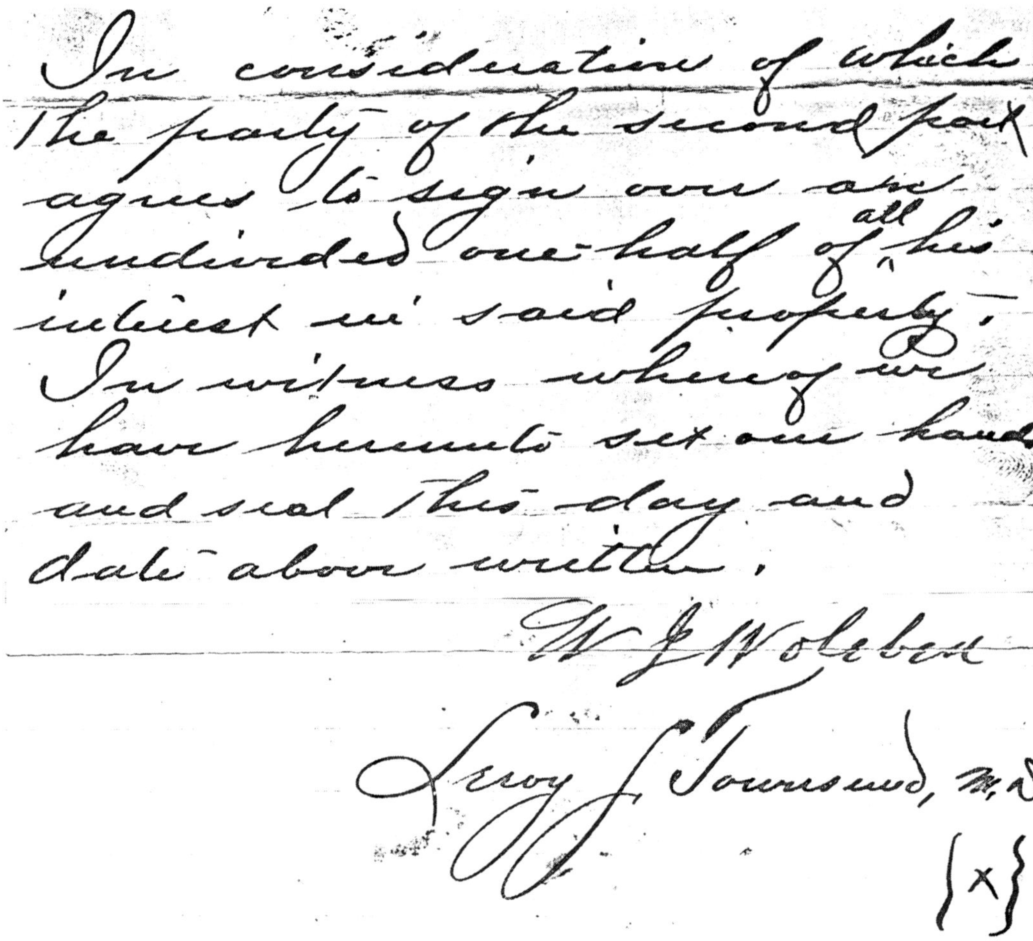

In consideration of which
The party of the second part
agrees to sign over an
undivided one-half of all his
interest in said property.
In witness whereof we
have hereunto set our hand
and seal this day and
date above written.

W J Woliben

Leroy S. Townsend, M.D.

{x}

Transcript of June 8, 1899 Agreement with W. J. Woliben

Port Valdez, Alaska.
June 8, 1899.

This agreement, made and entered into this eighth day of June, 1899, by and between W. J. Wiloben [he signs "Woliben"] of Lincoln, Neb., party of the first part, and Leroy S. Townsend, M. D., of Beaver Falls, Pa, party of the second part, Witnesseth: That the party of the first hereby agrees to do all necessary work to secure the sight [sic] and title of the party of the Second part to a certain puoficily [proficiency, ?], known as the Mul-ligau [?] Claims, and situated in the Teikel District, Alaska, subject to agreement—entered into by original locators.

[page 2]

In consideration of which the party of the second part agrees to sign over an undivided one-half of all his interest in said property. In witness whereof we have hereunto set our hands and seal this day and date above written.

W J Woliben
Leroy S. Townsend, M. D.

This last document reveals something of the future of Col. Creighton, the president and cohesive factor of the Keystone Company. This letter surfaced from Special Collections for manuscripts from the University of Washington. It reveals a verbal, if not loquacious, entrepreneur still heading up mining and land development ten years later. (Spelling, spacing, and punctuation are as they appear in the typewritten letter.)

GROUSE CREEK YOU LIKE MINING COMPANY.
OFFICE 623 "THE ROOKERY"

Spokane, Wash., March 6, 1909.

Hon. Richard A. Ballinger,
Secretary of the Interior,
Washington, D. D.

My Dear Judge:

Permit me to congratulate you upon the exhaulted position that you have so worthly attained.. With yourself in the Cabinet of Pres. Taft; the Hon. Wesley L. Jones in the Senate, and Judge Poindexter in his place together with the other three able gentlemen, this State may well be proud of its representation in the Council of the Nation. When you were appointed commissioner of the general lad office, you may remember, I told you I would much prefer to see you in Congress, but you have taken a short-cut and surpassed at one bound, Governors of states and Congressmen, to the great delight of all your friends and well wishers. From the time of your return from Washington with the hearty commendation of Pres. Roosevelt I for one felt sure that you would have a secretaryship in the next Cabinet, and frequently expresses my conviction to that effect. I have the utmost confidence, that in your case at least, the future will prove the wisdom of Pres. Taft's selection of Cabinet Ministers, and your State and your friends have additional grounds to be proud of the selection. Secretary Knox is a graduate of the same college that my brothers and myself attended. I am satisfied that no degree of success will ever cause you to lose interest in your former acquaintances and that you will be pleased to learn that the mining enterprise in which I am extensively interested at Wallace, Idaho, is turning out well.

Wishing you good health, happiness and an abiding success,

I am,

Very truly yours,

[signed in large, flowing hand:] Hugh R Creighton

[Reply]

March 15, 1909

My dear Colonel:

Yours of the 6th instant duly at hand. I thank you for the expressions of confidence and esteem expressed in your letter and also thank you for your congratulations and good wishes therein contained.

With best regards,

Yours very truly,

[Signed] R. A. Ballinger

Col. Hugh R. Creighton,
623, "The Rookery,"
Spokane, Washington.

Works Cited

Abercrombie, W. R. *Copper River Exploring Expedition, 1899.* Washington: Government Printing Office, 1900. Abercrombie's 1898 explorations are described on pp. 297-351. Appendixes A and B include sections from this publication.

---. "ALASKA.—1898. Copper River Exploring Expedition." *U. S. Compilations of Narratives: Explorations in Alaska 1869-1900.* Washington: Government Printing Office, 1900. 561-590.

Atwood, Fred N., ed. *The Alaska-Yukon Gold Book: A Roster.* Seattle: The Alaska Yukon Pioneers and The Ladies of the Golden North, 1930.

Baker, Marcus. *Geographic Dictionary of Alaska.* 2nd Ed. Prepared by James McCormick. Washington: Government Printing Office, 1906.

Beaver Star. Newspaper, Beaver Fall, Pennsylvania. Cited by Peggy Jean Townsend.

Berg, Henry C. and Edward H. Cobb. *Metalliferous Lode Deposits of Alaska.* USGS Bulletin 1246, Washington: Government Printing Office, 1967.

Burton, Pierre. *Klondike Fever: The Rush.* New York: Knopf, 1958.

Chicago Record. Klondike: Chicago Record's Book for Gold Seekers. Chicago: Monarch Book Co., 1897.

Christian Standard. Cited by Jim Cook.

Cobb, Edward H. *Placer Deposits of Alaska.* USGS Bulletin 1374. Washington: Government Printing Office, 1973.

Cohen, Stan. *A Klondike Centennial Scrapbook.* Missoula: Pictorial Histories Publishing Company, 1996.

Conger, Horace S. *In Search of Gold: The Alaska Journals of Horace S. Conger, 1898-1899.* Eds. Carolyn Jean Holeski and Marlene Conger. Anchorage: Alaska Geographic Society, 1983.

"Coroner's Inquest on the Deaths in Logan's Party." Port Valdez, Alaska, March 4, 1899. Copy of original typescript provided by Nancy Lethcoe. Appendix D.

Dawson City. Vol. 15, Nr. 22. Ed. Penny Rennick. Anchorage: Alaska Geographic Society. 1988.

Fortuine, Robert. *Chills and Fever: Health and Disease in the Early History of Alaska.* Fairbanks: University of Alaska Press, 1992.

Geographic Dictionary of Alaska (GDA). 2nd ed. Marcus Baker. United States Geological Survey, Department of the Interior. Washington Government Printing Office, 1906.

Gibbs, Jim. *Alaska Maritime.* Atglen, PA: Schiffer Publishing, 1997.

Glenn, Edwin F. and Capt. W. R. Abercrombie. Reports of *Explorations in Territory of Alaska, Cooks Inlet, Sushitna, Copper and Tanana Rivers, 1898.* U. S. Adjutant General's Office, Publication 25. Washington: Government Printing Office, 1899.

Globe. Newspaper, Beaver Fall, Pennsylvania. Cited by Peggy Jean Townsend.

Guiteau, Luther W. "A Plague of Scurvy." *The Alaska Sportsman* (January 1941): 16-18, 26-28.

Hanable, William S. *Alaska's Copper River: The 18th and l9th Centuries.* Anchorage: Alaska Historical Commission, 1982.

Haskell, William B. *Two Years in the Klondike and Alaskan Gold-Fields: 1896-1898.* Fairbanks: University of Alaska Press, 1998.

Holeski, Carolyn Jean and Marlene Conger, eds. *In Search of Gold: The Alaska Journals of Horace S. Conger, 1898-1899.* Anchorage: Alaska Geographic Society, 1983.

Lethcoe, Jim and Nancy. *Valdez Gold Rush Trails of 1898-99.* Valdez: Prince William Sound Books, 1996.

Margeson, Charles A. *Experiences of Gold Hunters in Alaska.* Self-published, 1899. Reprinted. Eds. Jim and Nancy Lethcoe. Valdez: Prince William Sound Books, 1997.

Mendenhall, Walter C. "Geology of the Central Copper River Region, Alaska." USGS Professional Paper No. 41. Washington: Government Printing Office, 1905. 20.

Neitzel, Michael C. *The Valencia Tragedy.* Surrey, B.C.: Heritage House, 1995.

News-Tribune. Newspaper, Beaver Fall, Pennsylvania. Cited by Peggy Jean Townsend.

Orth, Donald J. *Dictionary of Alaska Place Names.* Geological Survey Professional Paper 567. Washington: U. S. Government Printing Office, 1967.

Phillips, James Wendell. *Alaska-Yukon Place Names.* Seattle: U of Washington Press, 1973.

Remington, Charles. *A Golden Cross On Trails from the Valdez Glacier.* Los Angeles: White-Thompson, 1939.

Rohn, Oscar. *A Reconnaissance of the Chitina River and the Skolai Mountains, Alaska.* 21st Annual Report of the USGS, 1899-1900, Part II. Washington: Govt. Printing Office, 1900.

Seattle Post Intelligencer. Newspaper. Seattle, Washington.

Shipwrecks of the Alaska Shelf and Shore. OCS Report MMS92-0002. U.S. Department of the Interior Minerals Management Service. Alaska OCS Region. Managed by Mike Burwell. Internet: http://www.mms.gov./alaska/ships/index.htm. December 1998.

Walbran, Captain John T. *British Columbia Coast Names: Their Origin and History.* Vancouver, Canada: Douglas & McIntyre, 1909 and 1971.

Weekly Tribune. Newspaper. Beaver Falls, Pennsylvania. Cited by Peggy Jean Townsend.

INDEX and GLOSSARY*

*The resources for some descriptions of sites are the *Geographic Dictionary of Alaska* (GDA), 1906, 2nd ed. & the *Dictionary of Alaska Place Names* (Orth), 1966. Bold Face last names indicate persons active in Alaska.

About the Editors

Peggy Jean Townsend is a retired teacher of world and English literature, grammar and composition, and world cultures at high schools in Western Pennsylvania. She was born in Beaver Falls, Pennsylvania, and resides at Lime Kiln Hollow Farm, Darlington, PA. Of this locale, Peggy Jean writes, "The fields are green and rolling; the woods are full of Eastern hemlock, American beech, oaks, maples, and wildflowers; and our history goes back to 1834 when the farm was purchased." She and members of her family are at work on a variety of local history projects, one of which is Beaver County, Pennsylvania, history, which may be accessed on the Internet: www.basd.K12.pa.us.

Patricia Roppel is a forty-year resident of Alaska's Southeast panhandle; she has traveled, photographed, and written extensively about its culture and history. Her articles have appeared in Alaska magazines, and she has edited issues of *Alaska Geographic*. She has served on the Alaska Historical Commission and the Alaska Humanities Forum, and received various awards commending excellence in historical research. A colleague of Robert N. DeArmond, one of Alaska's foremost historians, Patricia Roppel kindly consented to assist in writing for and editing this work. Some publications by Patricia Roppel include *Alaska's Salmon Hatcheries, 1891-1959* (Anchorage: Alaska Pacific University Press, 1982) and *Southeast Alaska: A Pictorial History* (Virginia Beach: Donning Co., 1983). Due in 1999 is *LAND OF MISTS, Misty Fjords National Monument.* Projects include the economic history of Ketchikan and Alaska during prohibition.

Art Petersen has taught composition and literature at the University of Alaska Southeast in Juneau since 1975. In addition to volumes of poetry and academic texts, he has edited two books on the Alaska gold rush period: *Murder, Madness, and Mystery*: An Historical Narrative of Mollie Walsh Bartlett (co-authored with D. Scott Williams, Castle Peak Editions with Klondike Research, 1991) and *Gold Fever: A Narrative of the Great Klondike Gold Rush, 1897-1899* (Castle Peak with Klondike Research, 1997). Further research on Mollie Walsh-Bartlett, in the United Church Archives in Toronto, Canada, led to discovery of a manuscript about the Alaska/Canada Gold Rush by the Rev. Robert M. Dickey, the first "man of the cloth" into Skagway and the Klondike during that period. Petersen worked with the descendants of Rev. Dickey in Ottawa, Canada, on the manuscript and with his many photographs and papers. Mollie Walsh also led to contact with Peggy Jean Townsend and work on this book.

The editors acknowledge the generous contributions of Jim W. Cook of North Pole, Alaska. While writing a dissertation titled "The History of the Churches of Christ, Christian Churches and Disciples of Christ in Alaska," Jim learned of the Townsend letters from the Valdez Historical Museum and Archives and requested a copy of the working manuscript. Within days of receipt, Jim forwarded several pages of notes, including generally helpful information and full names of various members of the Keystone Company, which he had been researching. In response to a further draft of the manuscript sent to Jim, we received a letter from Cleta Cook, Jim's wife, informing us that he had "died peacefully in his sleep" on May 19, 1998, from a heart attack. "He was 45 years old. He was to have completed his PhD this December." We lament the loss of a generous colleague and husband.

R to L: Peggy Jean Townsend, editor, and Art Petersen, general editor, meeting over the Leroy Townsend letters, papers, and photographs at Lime Kiln Hollow Farm, PA, July 1995. (T. Pasteris)